Food Plant Engineering Systems

SECOND EDITION

Food Plant Engineering Systems

SECOND EDITION

T.C. Robberts

CRC Press
Taylor & Francis Group
Boca Raton   London   New York

CRC Press is an imprint of the
Taylor & Francis Group, an **informa** business

CRC Press
Taylor & Francis Group
6000 Broken Sound Parkway NW, Suite 300
Boca Raton, FL 33487-2742

First issued in paperback 2016

© 2013 by Taylor & Francis Group, LLC
CRC Press is an imprint of Taylor & Francis Group, an Informa business

No claim to original U.S. Government works

Version Date: 20121026

ISBN 13: 978-1-138-19939-2 (pbk)
ISBN 13: 978-1-4398-4809-8 (hbk)

Library of Congress Cataloging-in-Publication Data

Robberts, Theunis C.
    Food plant engineering systems / author, Theunis Christoffel Robberts. -- Second edition.
       pages cm
    Includes bibliographical references and index.
    ISBN 978-1-4398-4809-8 (hardcover : acid-free paper)
    1. Food processing plants--Equipment and supplies. I. Title.

TP373.R55 2013
664'.00682--dc23

2012034163

Visit the Taylor & Francis Web site at
http://www.taylorandfrancis.com

and the CRC Press Web site at
http://www.crcpress.com

To my two granddaughters, Mia and Keira, who believe in me as a teacher and inspire me to do a bit more, a bit better.

Contents

Preface xvii

The Author xxi

1 The System Approach to Lean Manufacturing 1
  1.1   Manufacturing and Trade 1
  1.2   The System 3
       1.2.1   Mass Production—The Auto Industry in Perspective 3
       1.2.2   Decline of Manufacturing 4
  1.3   Lean Thinking and Lean Production 5
       1.3.1   Lean Business Measurement Tools 9
              1.3.1.1   Waste Elimination as a Step toward Lean
                          Manufacturing 9
              1.3.1.2   The Seven Wastes 10
              1.3.1.3   Committing to Lean Production 11
              1.3.1.4   Implementing a Lean Program 12
              1.3.1.5   Lean Production Implementation Steps 13
              1.3.1.6   Implementation Strategy 14
       1.3.2   Quantifying the Value of Lean Manufacturing 16
              1.3.2.1   Lean Accounting 16
       1.3.3   The Skinny on Lean Implementation 17
              1.3.3.1   Strategy to Move Lean Thinking Forward 19
       Questions 21
       References 21

2 Measurements and Numbers 25
  2.1   History 26
  2.2   Standards 27
  2.3   Dimensions 28
  2.4   Units 28
       2.4.1   Base Units 29
       2.4.2   Definitions 29
       2.4.3   Supplementary Units 31

        2.4.4   Derived Units 31
  2.5   Dimensional Analysis 32
  2.6   Units for Mechanical Systems.. 34
  2.7   Conversions and Dimensional Analysis 37
  Problems 38

3   **General Calculations..41**
  3.1   Basic Principles 41
  3.2   Conservation of Mass 42
        3.2.1   Mass Balances.. 42
  Problems 53

4   **Properties of Fluids 57**
  4.1   Physical Properties of Flow 57
        4.1.1   Pressure 57
        4.1.2   Density (ρ) 59
        4.1.3   Specific Mass (w) 59
        4.1.4   Specific Gravity (SG) 59
        4.1.5   Bulk Modulus (k) 59
        4.1.6   Ideal Fluid 60
        4.1.7   Viscosity 60
        4.1.8   Temperature Effects on Fluid Flow................................62
        4.1.9   Shear-Thinning Fluids 62
                4.1.9.1   Bingham Plastics 62
                4.1.9.2   Pseudoplastics................................63
        4.1.10 Shear-Thickening Materials 63
  4.2   Surface Tension and Capillary Action 64
        4.2.1   Vapor Pressure 66
        4.2.2   Atmospheric Pressure.. 66
  4.3   Hydrostatics 67
        4.3.1   Pressure Intensity..67
  4.4   Basic Concept of Fluid Motion..69
        4.4.1   Introduction 69
        4.4.2   Conservation of Mass 70
  4.5   Types of Flow 71
        4.5.1   Turbulent and Laminar Flow................................71
                 4.5.1.1   Reynolds Number 72
        4.5.2   Rotational and Nonrotational Flow................................74
        4.5.3   Steady and Unsteady Flow................................75
        4.5.4   Uniform and Nonuniform Flow 75
  4.6   Conservation of Energy in Liquids 75
        4.6.1   Potential Head..76
        4.6.2   Pressure Head..76

4.6.3 Velocity Head 76
4.6.4 Energy Changes and Flow 77
4.6.5 Frictional Losses 78
    4.6.5.1 Frictional Losses in Straight Pipes 78
    4.6.5.2 Fittings in Pipes 80
4.6.6 Mechanical Energy Requirement............................81
Problems 82

5 Pumps..85
5.1 Introduction 85
5.2 General Principles...85
5.3 Pump Efficiency 86
5.4 Centrifugal Pumps 87
    5.4.1 Operation 89
        5.4.1.1 Net Positive Suction Head 91
5.5 Suction Systems and Sump Design.............................91
5.6 Positive Displacement Pumps 94
    5.6.1 Rotary Pumps......................................97
        5.6.1.1 External Gear and Lobe Pumps...................97
5.7 Pipelines in the Industry..................................102
    5.7.1 Piping Materials 102
        5.7.1.1 Pipe Assembly.................................103
    5.7.2 Stainless Steel Pipeline...............................104
    5.7.3 Welding of Sanitary Pipe 104
5.8 Pipeline Construction......................................104
Problems 106

6 Thermodynamics 109
6.1 Energy of a System 109
6.2 Characteristics of a System 110
    6.2.1 Energy of a System 110
6.3 Gases and Vapors..112
    6.3.1 Ideal Gas Equation 113
    6.3.2 Pressure Expressions 113
        6.3.2.1 The Perfect Gas Law 115
    6.3.3 Mixtures of Gases 117
Problems 120

7 Electrical Systems 121
7.1 Electricity Generation......................................121
7.2 Definitions 122
7.3 Energy Conversion 123
    7.3.1 Electrical Energy 125
7.4 Electric Motors 125

      7.4.1   Single-Phase Motors 125

              7.4.1.1   Single-Phase Universal Motor..........................125

              7.4.1.2   Single-Phase Split-Phase Motor 125

              7.4.1.3   Single-Phase Capacitor Motor 126

      7.4.2   Three-Phase Squirrel-Cage Induction Motors..................126

  7.5   Motor Management..128

  7.6   Power Transmission...128

      7.6.1   Flat Belts...129

      7.6.2   V-Belts..129

      7.6.3   Gears...130

      7.6.4   Calculation of Shaft RPM 131

      7.6.5   Management of Drive Systems 132

  7.7   Control Equipment for Electrical Apparatus 132

  7.8   Illumination of the Processing Facility 133

      7.8.1   Light Intensity and Application 134

      7.8.2   Types of Lamps 134

8   Heating Systems for Processing Plants................................137

  8.1   Heat Transfer 137

  8.2   Conductive Heat Transfer 137

      8.2.1   Resistance to Heat Flow 139

      8.2.2   Heat Flow through Cylindrical Walls....................141

              8.2.2.1   Steady-State Conduction through a

                          Composite Cylinder...........................142

  8.3   Convective Heat Transfer 143

      8.3.1   Heat Transfer by Conduction and Convection 144

  8.4   Heat Transfer by Radiation 146

      8.4.1   Stefan's Law..146

  8.5   Steam Heating..148

      8.5.1   Enthalpy...148

      8.5.2   Energy 149

      8.5.3   Specific Heat Capacity...150

      8.5.4   Specific Latent Heat..150

      8.5.5   Energy Associated with Steam 150

  8.6   Heat Pressure Diagram for Water 151

      8.6.1   Properties of Steam..152

              8.6.1.1   Saturated Liquid................................152

              8.6.1.2   Saturated Vapor.................................153

              8.6.1.3   Wet Steam..153

              8.6.1.4   Superheated Steam 154

  8.7   Steam Tables...154

	8.8		Boiling Point of Water		156
		8.8.1	Heating with Live Steam		156
		8.8.2	Culinary Steam		156
	8.9		Hot Water Systems		157
		8.9.1	Energy Calculations for Water Heating		157
	8.10		Piping		158
		8.10.1	Expansion of Pipes		159
	Problems				160
9	**Steam Generation**				**163**
	9.1		Boilers		163
	9.2		Fire Tube Boilers		164
	9.3		Water Tube Boilers		165
	9.4		Boiler Efficiency		166
		9.4.1	Fuel-to-Steam Efficiency		166
			9.4.1.1	Combustion Efficiency	166
			9.4.1.2	Thermal Efficiency	167
			9.4.1.3	Stack Temperature	167
			9.4.1.4	Radiation Loss	168
		9.4.2	Boiler Blowdown		168
		9.4.3	Water Treatment		169
	9.5		Boiler Selection Criteria		169
		9.5.1	Codes and Standards		170
		9.5.2	Steam Pressure		170
		9.5.3	System Load		171
			9.5.3.1	Heating Load	171
			9.5.3.2	Process Load	171
			9.5.3.3	Combination Load	171
			9.5.3.4	Load Variations	172
			9.5.3.5	Load Tracking	172
		9.5.4	Number of Boilers		172
			9.5.4.1	Downtime Cost	173
		9.5.5	Boiler Turndown		173
		9.5.6	Fuels		174
			9.5.6.1	Fuel Oil	174
			9.5.6.2	Gaseous Fuels	175
			9.5.6.3	Solid Fuels	176
			9.5.6.4	Stack Emissions	176
		9.5.7	Replacement Boilers		177
	9.6		Boiler Fittings and Accessories		178
		9.6.1	Pressure Gauge		178
		9.6.2	Steam Pressure Regulating Valve		178

9.6.3 Steam Trap 178
9.6.4 Water Column Tube 179
9.6.5 Water Injection 180
9.6.6 Safety Valve 180
9.6.7 Fusible Plug 182
9.7 Steam Purity 182

10 Refrigeration and Freezing 183
10.1 Overview 183
10.2 Natural Refrigeration 184
    10.2.1 Natural Refrigeration to Subzero Temperatures 185
10.3 Solid CO_2 185
10.4 Mechanical Refrigeration 185
    10.4.1 Capacity 187
    10.4.2 The Evaporating Coil 187
        10.4.2.1 Removal of Oil 188
    10.4.3 Expansion Valve 188
    10.4.4 Back-Pressure Regulator 189
    10.4.5 Compressors 189
    10.4.6 The Condenser 191
    10.4.7 Automatic Control 191
    10.4.8 Refrigerants 192
        10.4.8.1 Ammonia 193
        10.4.8.2 Other Refrigerants 193
10.5 Thermodynamics of Vapor Compression 196
    10.5.1 Coefficient of Performance 199
10.6 Secondary Refrigeration 201
    10.6.1 Congealing Tank System 201
    10.6.2 Brine System 202
10.7 Management 204
10.8 Storage Rooms 205
    10.8.1 Insulation 206
    10.8.2 Prefabricated Polyurethane Panels for Cold Room
        Construction 206
    10.8.3 Cold Room Doors and Devices 208
    10.8.4 Management of Cold Rooms 209
        10.8.4.1 Defrosting 210

11 Water and Waste Systems 211
11.1 Water Quality 211
11.2 Potable Water 212
    11.2.1 Water Safety 212

11.2.1.1 Health Concerns 213
11.2.1.2 Aesthetic Concerns 215
11.3 Substances in Water..215
    11.3.1 Hardness 215
    11.3.2 Iron 216
    11.3.3 Manganese 217
    11.3.4 Nitrates..217
    11.3.5 Chlorides and Sulfates 217
    11.3.6 Gases 217
    11.3.7 Silica..218
11.4 Treatment of Water Supplies...218
    11.4.1 Treatment to Remove Turbidity 218
    11.4.2 Softening Hard Water 219
        11.4.2.1 Cold Lime Method............................ 220
        11.4.2.2 Ion Exchange..................................... 220
    11.4.3 Water Demineralizing 222
    11.4.4 Filtering.. 222
    11.4.5 Care of Filters and Ion Exchangers 223
11.5 Treatment of Boiler Feedwater..224
    11.5.1 Deaeration 224
11.6 Wastewater Treatment 225
    11.6.1 Water Conservation 226
    11.6.2 Fluming Debris 227
11.7 Treatment Facilities 227
    11.7.1 Aerated Lagoon 228
    11.7.2 Trickling Filters 228
    11.7.3 Activated Sludge 228
    11.7.4 Anaerobic Treatment 228
11.8 Waste Disposal 229
    11.8.1 Prevention of Waste..229
    11.8.2 Solid Wastes 230
    11.8.3 Waste Incineration..230
        11.8.3.1 Controlled Incineration 231
        11.8.3.2 Incinerator Emissions 231
    11.8.4 Landfill...231
        11.8.4.1 Landfill Methods...............................232
        11.8.4.2 Public Health and Environment 232

12 Materials Handling 235
12.1 Importance of Materials Handling 235
12.2 Increasing Materials Handling Efficiency 236
    12.2.1 Designing a Materials Handling System.....................237

   12.2.2 Kinds of Systems 238
   12.2.3 Analysis of a Materials Handling System.........................238
  12.3 Materials Handling Equipment.......................................241
   12.3.1 Conveyors...241
    12.3.1.1 Chutes 241
    12.3.1.2 Roller Conveyors 244
    12.3.1.3 Belt Conveyors 244
    12.3.1.4 Slat Conveyors..................................245
    12.3.1.5 Chain Conveyors...............................245
    12.3.1.6 Vibratory Conveyors.........................245
    12.3.1.7 Screw Conveyors 245
    12.3.1.8 Bucket Elevators 246
   12.3.2 Trucks 246
   12.3.3 Pallets 247
   12.3.4 Bulk Handling 247
  12.4 Plant Layout for Materials Handling...............................248
   12.4.1 Space Arrangements 248
  12.5 Efficient Use of Materials Handling Equipment 249
   12.5.1 Maintenance..249

13 **Manufacturing Plant Design..251**
  13.1 Building Design 251
  13.2 Legal Aspects...251
   13.2.1 Building Bylaws..252
   13.2.2 Sanitation Codes 252
   13.2.3 OSHA 252
   13.2.4 EPA 252
  13.3 Expansion...252
  13.4 Plant Location 253
  13.5 Planning the Building on the Site....................................256
  13.6 The Structure..257
   13.6.1 Foundations..258
   13.6.2 Supporting Structure..258
   13.6.3 Walls 259
   13.6.4 Floors...259
    13.6.4.1 Slope of Floors..................................260
    13.6.4.2 Kinds of Floors 260
    13.6.4.3 Drains 262
   13.6.5 Windows 263
   13.6.6 Doors 263
   13.6.7 Piping..263
   13.6.8 Electrical 264

13.6.9 Ventilation 264
13.6.10 Hand-Cleaning Stations 264
13.7 Facilities Layout 265
13.8 Summary 265

14 Environmental Issues 267
14.1 Costs and Benefits of Environmental Compliance 267
14.2 Mandated Environmental Costs 267
14.2.1 Change 268
14.3 Corporate Social Responsibility 268
14.4 Legislative Issues 270
14.5 Environmental Compliance of Operations 271
14.5.1 Introduction 271
14.5.2 Organization and Staffing for Environmental
Compliance 271
14.5.3 Strategic Management 272
14.5.4 Organizational Structure 272
14.5.5 Environmental Audits and Review 273
14.5.5.1 Solid Waste 273
14.5.5.2 Recycling 274
14.5.5.3 Air 274
14.5.5.4 Water 274
14.6 ISO 14000 Series Environmental Management Standards 275
14.7 Corporate Responsibility 277
14.8 Application of the Principles 278
References 279

15 Safety 281
15.1 Workplace Safety 281
15.2 Causes of Accidents 282
15.3 Safety 283
15.3.1 Hazards in the Food Industry 284
15.3.2 Handling Materials 284
15.4 Accident Prevention 284
15.4.1 Color Coding 285
15.4.2 Operating Instructions 286
15.4.3 Pressure Equipment 286
15.4.4 Grain Handling Equipment 287
15.4.5 Flammable Material 287
15.5 General Plant Safety 287
15.5.1 Inspection of Equipment 288
15.5.2 Power Drives 288

15.6 Fire Protection 289
    15.6.1 Fire Fighting 292
        15.6.1.1 Portable Extinguishers 292
15.7 Noise Control 293
15.8 Occupational Safety and Health Administration...........................295

Bibliography **297**

Appendix 1 ....**299**

Appendix 2 ....**305**

Appendix 3 ....**321**

Index **327**

Preface

It is important always to remember that we do business for the sake of making money. In a capitalist economic system, the *rules of business* are supposed to be kept to a minimum. The essence of capitalism is that competition shall ultimately provide the biggest benefit to all. All that restrains *totally free competition* are legislative attempts to make the competition as *fair* and *equal* as possible. The *First World* accepts capitalism in some or other guise as the *best* way to do business. The *essence* of capitalism is that all people are free to exchange, or barter, their goods to the best of their abilities. Presumably, this provides the maximum benefit for all to accrue wealth according to their abilities. In this paradigm the consumer has needs that must be met and the producer that matches these needs will have the most sales. Consumer requirements frequently change and manufacturing has to respond to these changes.

This book deals primarily with the ancillary equipment used in processing industries. Many books are dedicated to describing the equipment used in specific industries or general plants used in related industries, but many authors totally disregard the importance of ancillary equipment such as motors, pumps, pipes, boilers, valves, and controllers that allow the plant to operate. This book tries to fill the need for text dealing not only with the bits and pieces that keep systems running, but also with how the peripheral pieces of a processing plant fit within the bigger picture.

A production line needs to be assembled to manufacture a product or range of products. Such a line will include many subsystems that must merge into an integrated system. All parts of the system are important, and the obvious parts receive much attention. Some parts, such as the peripherals or services including pumps, boilers, power transmission, water treatment, waste disposal, efficient lighting, and many others, may be completely forgotten. These parts of the operation are as important as the actual processing machines; without them, the system will collapse. Any plant must conform to the system* approach.

---

* Defined by Webster's dictionary as "a regularly interacting or interdependent group of items forming a unified whole."

The information in this book is gathered from diverse sources and will be of particular value to plant operators, production managers, and students, who will gain understanding of the ancillary and peripheral components of the production line, environmental control systems, the building and facilities, and the way in which various parts of a plant interact to increase overall production, even though individual parts may operate at less than their optimal levels. This will lead to better understanding of the requirements for true plant optimization.

For maximum sales and optimum profits, manufacturers try to produce a commodity at the lowest possible cost, as close as possible to what consumers believe they require. Good marketing strategies can convince consumers that they require the goods that are produced, such as iPhones, iPads, and many other successful products. The balance between value for money and profit should be such that it leaves the consumer with a sense of "delight" about the purchase. The price that a consumer is willing to pay has nothing to do with the input cost. The objective of the manufacturing system is to produce the desired or required product with the least manpower, equipment, energy, lost motion, and, therefore, cost. Systems may be simple or complex. In most systems, there are one or more subsystems, all supposedly working together in proper sequence and each contributing to the total effort.

The automobile is a good example of multiple subsystems working together in a system. The fuel subsystem consists of a fuel tank, a fuel line, a pump, and the injector (or carburetor), which mixes the gasoline with air in the proper proportion to create an ideal explosive vapor. The explosion provides the pressure to push a piston that is connected to the rest of the power train. The power train includes the engine, clutch, transmission, driveshaft, differential, axle, and wheel. The elements mentioned may have different components that make the subsystem operate properly, but everything must work together efficiently to accomplish the desired outcome.

Engineers, line, and production managers should ensure that all the elements of the system are properly balanced to accomplish the job efficiently. If the driveshaft in an automobile were unable to deal with the torque, it would break and stop the operation, although all other elements would be in perfect working condition. On the other hand, if the element is too large, it could be unwieldy and inefficient. The same matter of judgment and proper design must be carried out on all of the machines, the building, and every part of the processing plant to obtain proper balance. The system components must work together to accomplish the required task, with each unit or subsystem functioning efficiently and dependably.

A successful system must provide for dependable operation. This is a function of continuous and preventative maintenance. Preventative maintenance requires data to predict when a component will fail. To optimize serviceability of electro-mechanical components prior to replacing them, predictive maintenance that will account for measured vibration monitoring, operating dynamic analysis, and infrared technology can be used. The primary focus of such a predictive maintenance system will be on the critical process systems. The operating system that produces the product that generates revenue is more important than the auxiliary equipment.

Getting all the units to work together as a well-orchestrated system is what design and management of manufacturing are all about. This starts with subsystem optimization that will eventually be combined in total optimization, which focuses on the bits and pieces of equipment that keep the plant going. Most of these pieces of equipment are taken for granted and are overlooked when systems are designed or maintenance schedules are written.

Manufacturing systems can be broken down into tasks or activities. Some of these tasks will be performed in sequence, while others will be carried out simultaneously. When a house is built, there is a sequence foundation, followed by walls, floors, and roof. Once the walls are done, the windows can be put in place while the furnace is installed in the basement. Any delay in a sequence of events will delay the project. A simultaneous activity that can be finished with time to spare will not delay the project if it is done within the allotted time. Once the slack or time available is at the point where a further delay will delay the project, all the slack has been used, and the activity becomes critical.

When a system is analyzed with the critical path method (CPM) or program evaluation and review technique (PERT), one can identify the activities or operations of the plant that have no slack. Keeping these critical elements optimally operational will ensure better balance in the system and optimized production. It is, however, a technique better taught in operations or project management. Thus, this book will describe the features and characteristics of the service systems that allow for implementation during line upgrading or replacement. Just remember that slack can disappear rapidly in the case of an unforeseen breakdown.

The System

Another objective of this book is to help the reader understand some of the terms frequently used regarding processing lines and give some of the principles underlying the terms. This book is written so that students and manufacturing personnel with only basic mathematical knowledge and a simple calculator will benefit from it. The examples were chosen to lead the reader to a full understanding of some of the mathematics to be expected in engineering texts.

Chapter 1 covers the systems approach to Lean manufacturing. In Chapter 2, measurements and numbers are described. Chapter 3 deals with general calculations, including mass and energy balances. Chapter 4 addresses the properties of fluids, which leads to Chapter 5 about pumps and piping, and finally to a brief discussion of thermodynamics in Chapter 6.

Chapter 7 deals with electrical systems, including motors, starters, electrical heating, and lights. Chapters 8 and 9 describes heating systems and steam generation, respectively. In Chapter 10, cooling and refrigeration systems are investigated.

Water and waste, and material handling systems, are covered in Chapters 11 and 12. Chapter 13 concentrates on site, foundations, floors, walls, roofs, drains, insulation, piping, and electrical systems. Chapter 14 focuses on environmental issues. An overview of safety and OSHA regulations is given in the final chapter. The appendices contain a number of conversion tables and an introduction to mathematics.

The Author

Theunis C. Robberts, PhD, is the program director for the manufacturing management majors at the University of Minnesota. He teaches principles of engineering, product development, quality systems, regulations and compliance, and business analytics.

He received his undergraduate qualification in food technology at Pretoria Technikon and his masters diploma in technology from Cape Technikon in South Africa. He is a candidate for a PhD in business administration, business quantitative methods, from Northcentral University in Prescott Valley, Arizona.

Dr. Robberts started his research in the production of diamond grinding wheels and later switched to doing sorghum beer research at the National Food Research Institute of the CSIR in South Africa. After that he started teaching and held various teaching positions at different polytechnics and universities. At the Technikon Witwatersrand, he built a large pilot plant in an old building on campus. This plant is still operating as a teaching and outreach facility. The retort, the exhaust box, and the bottom outlet jam-pot were built according to his designs. He also specified the requirements for the control instrumentation. The plant has a general-purpose canning line; processed meat line; milk pasteurizer and cheese bath; beer brewing and bottling line; baking facility, including mixers, molders, and deck ovens; a spray drier; a short screw extruder; and a section for product development.

One of his larger consulting projects took him to Mozambique where he upgraded a biscuit factory that had stagnated for 25 years. All the recipes for the stamped, rotary molded, and wire-cut biscuits were reformulated. The processing lines were upgraded. The Quality Control Department was updated and the instrumentation, including an infrared analyzer, alveograph, and various other laboratory equipment, was calibrated. The staff were trained to operate and maintain the equipment. They were also taught how to set up, maintain, and use statistical process control (SPC).

His present research and outreach are focused on the establishment of business parks and light manufacturing facilities.

Dr. Robberts received the Ernest Oppenheimer Memorial Trust Award for senior researchers. He has published articles and presented lectures at several scientific meetings, including the Institute of Food Technologists (IFT), where he is a professional member. He received the Distinguished Faculty Award from the University of Minnesota. The manufacturing management program received a citation for excellence in practice from the American Society for Training and Development (ASTD) for the outreach and teaching at Marvin Windows & Doors.

Chapter 1

The System Approach to Lean Manufacturing

1.1 Manufacturing and Trade

Manufacturing and service facilities exist to make money. Capitalism is based on trade and our lifestyle is based on a consumer that is able to purchase goods and services. One party sells, barters, or trades goods, commodities, or services to another party. Every change of ownership or rendering of service has concomitant financial rewards for the seller or service provider and potential real or assumed rewards for the receiver, buyer, or consumer.

The economy of most countries is based on revenue derived from manufacturing and service industries. Countries that enter the phase of economic upswing normally rely heavily on manufacturing to provide the commodities that are bartered. This frequently provides numerous positions for unskilled or semiskilled employees, who then become economically more active. They are able to purchase more goods and services that stimulate more manufacturing and service-related positions. As the economy matures and society reaches a higher level of education, more emphasis is placed on service industries and the economic contribution of the manufacturing industries becomes less important. At the stage when the economy relies mainly on service industries, the economy of that country reaches the "retirement" stage.

In the United States there is still some balance between manufacturing and service industries, indicating a mature but still expanding economy. The reason for the extended period of economic growth in the United States is the ability of manufacturers to retool their manufacturing operations to provide altered/new/different products that are customized for an ever-changing marketplace with novel needs. Service industries have been doing this for a long time. Service and manufacturing industries rely on an educational system that will provide them with skilled workers that are educated in a way to allow for fast retraining. This will allow employers to provide contemporaneous manufacturing or service that evolves with changing consumer requirements.

The manufacturing companies in the United States with the largest amount of slack are those companies that sell products with a very low intrinsic value and a high price. Think about the material and production cost to manufacture a Windows DVD and the selling price for that DVD. Most of the cost associated with that product is embedded in the people writing the code and those that yield after-sales service. The same is true for music CDs. These companies can increase production without any increase in factory or plant size. In the case of electronically delivered software, music, games, or books, the "product" that is downloaded becomes a service that is rendered by the vendor. The storage and distribution costs of electronic products are relatively small compared to costs for products with volume and weight.

Manufacture of commodities like food, automobiles, appliances, tools, and other stuff we buy in retail stores relies on assembly-style manufacturing where an increase in sales forces an increase in capacity. If the factory operates for 24 hours per day, the capacity is fully utilized and increased capacity will require large capital investments and major updating that may take several years to accomplish. At the time when the sales of the product stop growing or decline, then the product needs to be changed. There is normally a considerable lag time before a production system can be altered to respond to the changed marketplace. For this reason, products continuously evolve to satisfy new requirements so that the altered product can be produced on the same processing line. "New" products are frequently also regular products packaged in a different way. At the time when the KitchenAid mixer became available in new colors, a new product line was created by changing the color of the paint in the production line.

To manufacture a commodity, a production line needs to be assembled. Such a line will comprise many subsystems that will eventually operate as an integrated system. An intrinsic but frequently forgotten part of this system comprises the peripherals or services such as pumps, boilers, power transmission, water treatment, waste disposal, efficient lighting, and many others. Since this part of operation is as important as the actual processing machines, they will be treated using the system approach.

Analysis of the system with the critical path method (CPM) or program evaluation and review technique (PERT) will undoubtedly lead to better balance in the system and optimized production. It is, however, a technique better taught in operations or project management.

1.2 The System

Manufacturers try to produce a commodity as close as possible to what a consumer requires at the smallest possible cost. The objective of the manufacturing system is to produce the desired product with the least work force, equipment, energy, lost motion, and, thus, cost. Systems may be simple or very complex. Usually, in every system, there are one or more and sometimes dozens of subsystems, all supposedly working together in proper sequence and each contributing to the total effort.

Consumers are not concerned with production, but with the presence of the product in the retail venue. The other important part of the complete system will include the distribution of the product from the production site to the various storage facilities and finally on to the retail shelf. Also, the raw materials and components must be delivered to the production facility in time for the production to take place. Logistics and transportation systems are integral parts of the total system but are part of the larger operational plan. Getting all the different parts of a system to work together as an integrated system is what manufacturing is all about.

1.2.1 Mass Production—The Auto Industry in Perspective

The first Model A Ford left the factory on June 16, 1903. From the beginning, Henry Ford realized that manufacturing techniques could make his product affordable for the public. He focused his efforts on the manufacturing process to ensure maximum efficiency to reduce costs. In 1914, the first moving conveyor belt was introduced in Ford's Highland Park plant. This system increased production and decreased costs (Ford 2009).

Henry Ford was quick to realize that the burgeoning labor force in the US manufacturing industry would be able to afford his cost-effective cars. He pushed for more gas stations and campaigned for better roads. He also realized that planning should include the global market and sold his product in more than 30 countries before his competitors realized the opportunity of globalization (Ford 2009).

In the 1930s, mass production of automobiles in the United States was the norm. At that stage, the Japanese market was small when compared to America and Europe. The management at Toyota studied Ford's production system, where a limited number of models were produced in high volumes. At that time, Toyota produced a low volume of a large number of models on a single line (Wu et al. 2007).

During the 1950s, the Toyota management team undertook a 12-week study tour of the US automobile industry. The team understood the benefits of mass production and the inherent problems such as high inventory caused by large batches, detached process steps without time matching, high equipment cost, overproduction of models, hidden defects, and quality problems. The result of this study allowed the management of Toyota to formulate their own system that pursued quality, with standardized processes, continuous material flow, and waste elimination. Toyota also changed inventory from a push to a pull system, and designed the *kanban* system (Wu et al. 2007).

This new system helped Toyota to gain market share and, during the 1970s, it became the number six automaker. In 1994, Toyota replaced Chrysler as the number three automaker. In 2003, Toyota passed Ford in global market sales. At the end of the fiscal year in March 2003, Toyota's annual profit was $8.13 billion and this profit was larger than the combined profits of General Motors, Ford, and Chrysler. Toyota's net profit margin was 8.3 times higher than the average margin for the industry. Toyota's market capitalization at $105 billion in 2003 was also higher than the combined market capitalization of General Motors, Ford, and Chrysler (Wu et al. 2007).

1.2.2 Decline of Manufacturing

Manufacturing in the United States declined drastically over the last few decades as companies moved their manufacturing to foreign countries. The reasons for the outsourcing of manufacturing include the cost of local labor, the cost of occupational safety, various liability issues and concerns, environmental agency regulations, insurance costs, and very cheap labor and few or almost no regulations in other countries. This study will address the single issue of production costs.

The most obvious consequence of the loss of manufacturing jobs is unemployment of large numbers of people that were trained to do specific duties on a production line. The skills of these workers are not readily negotiable in other sectors of the economy, so for all intents and purposes the workers require retraining or will need to do unskilled labor at much lower salaries. To compound the problem, there are limited job opportunities for unskilled laborers since these jobs were the first to be outsourced to foreign manufacturing. The days of numerous laborers working side by side in a production setting are gone. Automation and sophisticated design-for-manufacturing techniques replaced many workers with a few workers. This is leaving the United States in a socioeconomic position that has a dire prognosis. When a parent can no longer provide for his or her family, the pride is taken from that individual and is replaced by despair. The loss of the home and many other items that are taken for granted becomes a new reality in the lives of many people in specific neighborhoods. Abuse of alcohol, drugs, and all that is worst in society becomes commonplace and, where we had hard working, tax-paying citizens, we now have desperate people on welfare. This new reality is hopefully transient and will be replaced with something better.

Entrepreneurial ventures that are too small to afford robotics and automation yet robust enough to grow both output and jobs are able also to grow employment opportunities. To ensure that these ventures are not only attractive to venture capitalists and other funding sources, but also are economically sustainable for many years, production will have to be done in a way that will allow maximum profit margin per input cost at the highest possible productivity. The way out of the recession is to ensure that each worker that wants to work will be able to find a job close to where he or she wants to work, at a salary that can sustain the family lifestyle that workers are used to in their communities.

With all the free-trade agreements, it is impossible for the federal government to subsidize manufacturing. That would have been the fastest and cheapest way out of the present recession. Lean manufacturing is one of the keys to sustainable high employment entrepreneurial ventures. Alcoa applied Lean production and saved $1.1 billion in the 2 years from 1998 to 2000 while improving the safety, productivity, and quality of their production (Spear 2004).

1.3 Lean Thinking and Lean Production

The basic principles of Lean are founded in waste reduction and efficiency optimization. These principles have been adopted by many businesses after they noticed the performance of the Toyota production system. Toyota uses value stream mapping to analyze the total input for a product from raw materials to the finished state. Value stream mapping is also used to analyze service and administrative processes according to the needs of the consumer until the service is performed to the customer's satisfaction (Mehta 2009; Baines et al. 2006).

To stay competitive in a global market, manufacturers must produce their goods at the lowest possible cost. The minimum wage in the United States is higher than the average wage paid in developing countries; local manufacturers have to compete with the adopted the concept of minimum wage and there is no way to reduce cost by employing cheap labor. In 2007, the average civilian worker earned $19.62 per hour and received benefits that cost the employer $28.11 per hour (Flores 2008). In Vietnam the minimum wage was increased in 2008 to $72 per month for people working in foreign invested companies in major cities. Locally owned companies have to pay a minimum wage of $48 per month (Conti 2009). When average workers work 160 hours per month, they will earn $0.30 to $0.45 per hour. At the current US Department of Labor's (2009) minimum wage of $7.25 per hour, local manufacturing companies spend 16 times more on an hour of labor than the company that manufactures its goods in a big city in Vietnam.

To produce anything in the United States with a high federally mandated minimum wage, the input costs must be reduced. People purchase goods or services to which value has been added. Any value that is created by an organization is the end result of numerous steps in the value-adding stream. The most important key to the survival of the organization is to get the right value in the right time and at the right price to the consumer. This requires that the steps in the value stream must be performed in the optimum sequence at the right time (Womack 2007).

In most manufacturing organizations, each part of the value stream has associated financial metrics to grade performance. This is done to get the best utilization of assets in that specific subsection of the total operation. This lack of integrating the whole value stream into a seamless value-creating effort causes the imbalance and wasteful production models that frequently exist within manufacturing systems. The way to ensure Lean production is to have someone responsible to oversee the value flow from initial input all the way to the customer. The flow must be continually improved in a way that will keep consumers delighted while making money for the enterprise. Each supervisor will know what every person must do to ensure the optimization of the value flow. Each worker must be allowed to contribute to the development of better ways to do business such that value addition will increase in the least possible time (Womack 2007).

Lean can be summarized as an effort to create more value with less work. Lean is a holistic approach that is used to eliminate costs that do not contribute value to the product or service that the consumer pays for. Lean thinking incorporates all aspects of operations with the sole aim of producing more saleable goods or services with less input. In the case of services that a consumer may find at a dentist, any resource that is idle is lost, because the resource time cannot be placed in inventory for later use (Schroeder 2008). If the dentist is not servicing someone, that service time is lost forever. Lean in this case would be to utilize the resource—dentist time—in the best possible way so that as little time as possible is lost.

In the narrow sense, a manufacturing operation is the process that transforms material to a higher level of value. Each operation thus imparts more value until the final product or service is delivered (Schroeder 2008; Krajewski, Ritzman, and Malhotra 2007). The four key elements in the operations decision-making process will impact the process, the quality, the capacity, and the inventory. Increased productivity is derived from all four of these elements (Schroeder 2008).

Within a manufacturing environment, not all of the unused resources are lost—for example, in the case of production time, time that is lost since time cannot be inventoried without prior arrangement. In the case of materials that are not used, it can be placed in raw material or semiprocessed inventory for later use. There is, however, cost associated with inventory such as insurance, handling, obsolescence, pilfering, and opportunity costs of the capital tied up by the cost of the goods in inventory. Keeping inventory to the absolute minimum is a primary aim of Lean thinking. Inventory is not always housed in a warehouse. When there are large numbers of items in the production line, a large amount of materials in process generates inventory that is tied up in production. Minimizing the production or throughput time will minimize this inventory and add to the profit (Schroeder 2008).

When the process is isolated from the rest of the operation it becomes apparent that a process that will do more with the same input or do the same with a smaller input will increase productivity or decrease the cost of production. At the same time, if the number of rejects is diminished, the number of salable products increase and productivity is increased. When capacity increases, there are concomitant increases in the economies of scale that cause increased productivity. Lastly, a decrease in inventory will decrease the cost of holding the inventory (Schroeder 2008).

Lean operations are driven by the production quality. When a factory produces 5 percent defective products that cannot be sold, the cost is 5 percent of the total input costs plus the loss of opportunity to sell the products. If the process can be enhanced to produce only 1 percent defective products, the daily output increases and losses decrease by 80 percent. Making things right the first time is thus an opportunity to reduce waste and increase production at no additional cost except for the cost associated with the quality assurance process. This cost is normally much lower than the costs associated with the loss of production.

Another aspect associated with Lean thinking related to quality is the avoidance of product failure after purchase. This includes costs associated with product recalls and the very expensive marketing campaigns that are required to build consumer confidence in the brand after a publicized national recall. There has been a steady rise in annual automobile recalls in the United States and almost 19 million vehicles were recalled in 2002. In 1988, 6 million pounds of meat and poultry were recalled. In 2003 about 26 million pounds of meat and poultry were recalled (Shang and Hooker 2005).

By one calculation, the direct cost to auto manufacturers with a recall is about $100 per vehicle. Excluding any indirect costs, automakers spent around $3 billion in 2004 to recall and fix noncompliance and safety defects in their products. The indirect costs associated with a recall include loss in goodwill or reputation. The brand is thus damaged as well as the marketing costs associated with this problem. It has been calculated that companies experience a $42.8 million average loss in shareholder value after a recall announcement (McDonald 2009).

For the past 140 years the United States had rapid economic growth because of the growth of industries. The industries provided jobs, and more jobs were created than could be filled by local labor; for that reason, the United States promoted an aggressive immigration drive. The influx of mainly manual labor sustained the growth of the manufacturing industry. A growing labor force of immigrants was encouraged to come to the United States through the open frontier program.

After the Civil War the open frontier was the center point of the US economic development until it reached its geographical limit in the 1890s. The final surge in railroad building occurred between 1886 and 1892 and this limited the possibilities for resource-based growth (Greasley and Oxley 1996). At this point, the growth of the economy was dependent upon manufacturing rather than expansion of settled territories.

During the 1890s manufacturers in the United States became competitive in the world markets. The per-capita production soared and exceeded that of Great Britain. The industrial leadership is attributed to technological development—in particular, electric power, the petrol engine, and an organizational structure that was set up to be suitable for mass marketing. The United States emerged through the better part of the twentieth century as the industrial giant. The oil shock of 1973 exposed an inherent fragility in industrial economies around the world. This exogenous force started the US economic slowdown (Greasley and Oxley 1996).

The US economy did not develop new products and technology as fast as they were developed in Japan and other newly industrialized economies, causing the first current account deficits since the 1890s. It is noteworthy that growth of labor productivity was not influenced by the oil shock, although workers were transferred to sectors of manufacturing that were less exposed to international competition (Greasley and Oxley 1996).

During the period from 1980 to 2000 the US economy witnessed a widening gap between skilled and unskilled wages. This can be explained by outsourcing of manufacturing and the concomitant decrease in manufacturing employment. At the same time, capital accumulation and expansion took place in nonmanufacturing employment. This shift in employment between the manufacturing sector and the nonmanufacturing sector intensified the wage inequality between unskilled and skilled workers. Because more skilled workers work upstream in the production side, outsourcing of service-related positions increases the wages of skilled workers and the relative demand for skilled workers (Chongvilaivan, Hur, and Riyanto 2009).

Since companies in the United States are outsourcing upstream production activities, we are left with downstream production activities and services. These activities are less skill intensive and fewer skilled employees are required.

1.3.1 Lean Business Measurement Tools

The Toyota system of removing all steps that do not add value, combined with the fourteen points of management promulgated by Dr. W. Edwards Deming to increase quality, forms the foundation of the Lean philosophy (Serrano, Ochoa, and De Castro 2008). When Lean strategies are pursued in manufacturing, it is imperative that measurement of performance should also become Lean. One such tool is Lean accounting that utilizes cost accounting according to value stream associations. Financial reports are also formulated to include nonfinancial information that is imperative to the understanding of the Lean benefits (Brosnahan 2008).

To enable value stream management of operations in a company, the company must be organized into value streams rather than into traditional departments such as customer service, purchasing, and so forth. All the functions and people that are needed to support the operations of the value stream are included in that value stream. The order fulfillment value stream includes all the functions and people, starting with sales and order entry through operations and manufacturing to shipping and after-sales service. One person is responsible for the profitability of the entire value stream and all the people in the value stream report to that person. The operational metrics that are monitored in a value stream include safety, quality, delivery, costs, and time (Brosnahan 2008).

Activity efficiency is used in manufacturing and service operations to balance processes and to let operators know how well they work as a team. In manufacturing and service operations, throughput and teamwork are essential. With value stream mapping, people that implement Lean principles calculate the value-added time percentage and the inventory efficiency for manufacturing. The percentage of value-added time is used as a metric with inventory efficiency (Mehta 2009).

It is important to note that most processes will start with an activity efficiency of about 62 percent of the operators' time to perform value-added work that will satisfy the consumers' needs. When the process is optimized for Lean operation, the activity efficiency can be increased to 82 percent. The 20 percentage point increase represents a real increase in activity efficiency of 32 percent based on the standard efficiency. This efficiency and therefore additional productivity comes without additional costs. It should also be noted that the activity efficiency is optimized at 82 percent. This 18 percent loss from the ideal 100 percent is caused by inherent imbalance within operations (Mehta 2009).

1.3.1.1 Waste Elimination as a Step toward Lean Manufacturing

The Japanese term used for hidden production waste that can be eliminated is *muda*. Within the Lean paradigm, waste is defined as any activity that utilizes resources but creates no value (Wood 2004).

Waste is invisible and difficult to eliminate. Within Lean manufacturing, a set of techniques used to identify and eliminate waste has evolved. The techniques include cellular manufacturing, pull scheduling (kanban), Six Sigma and total quality, single digit setup time, and team development (Lee 2008).

To eliminate waste, Wood (2004) uses five principles:

- The first principle relates to what the customer wants. The value of each product must be specified by the rule of needs.
- The second principle relates to identifying the value stream starting with product design, development, procurement, manufacturing, distribution, and sales in relation to the needs of customers. The functionality and capability of the supply chain must be scrutinized to detect and eliminate wastes.
- The third principle addresses issues of production flow. Bottlenecks should be eliminated and flow should be changed to single piece flow produced as close as possible to the rate of sale.
- The fourth principle addresses the elimination of inventory. Production is aligned to the needs of customers. Only produce what the customer requires, when the customer needs it.
- The fifth principle is the pursuit of perfection by complete elimination of waste. At this point, every activity creates value for the customer.

The "five principles" is a journey of continuous improvement with no tolerance to maintain a suboptimized status quo.

1.3.1.2 The Seven Wastes

Shingo (1989) identified seven types of manufacturing wastes:

1. Overproduction
2. Waiting time
3. Transport
4. Inventory
5. Motion
6. Defects
7. Overprocessing

The first waste is *overproduction*—when goods are produced in a quantity greater than required or earlier than necessary. This waste produces inventories of products, components, subassemblies, unfinished products, or even garbage when materials deteriorate during storage. The second waste relates to *waiting time* or delay. When employees wait for a machine to complete a process, their time is wasted. This delay can also be caused by a large amount of work in process because of large batch production, downstream equipment malfunction, or defective items requiring rework. Ultimately, this leads to increased lead time (Dennis 2007).

Transport is the waste associated with inefficient workplace layout, traditional batch layout, and excessively large equipment that forces a specific workplace layout. Transport between cells is an essential waste, while unnecessary movement is a correctable waste. *Inventory* is the waste that occurs when a logistics activity causes more inventory than what is required to be positioned in a location. This can be caused by early delivery or delivery to the wrong manufacturing center. Reduction of inventory and increased operational efficiency are achieved by reducing lead time and variation in demand. Excess inventory also occurs when production is scheduled according to a materials requirement planning system without recognition of the amount of goods that are sold (Dennis 2007).

Motion is waste caused by unnecessary movement for operators such as walking, reaching, and stretching. Wasted human motion is related to poor ergonomic design that negatively affects productivity, safety, and quality. Waste of machine motion exists when the machines in subsequent operations are not set close enough to each other to allow movement of the work piece from one operation to next operation (Dennis 2007).

Defects are very obvious and costly forms of waste and include any activity that causes rework or unnecessary adjustments or returns, such as product recalls, invoice errors, inventory discrepancies, mislabeled products, or any other error. The waste in this case comprises all the initial costs associated with the product and all subsequent costs to rectify the situation. The costs associated with dissatisfied and angry customers should also be calculated in renewed marketing efforts to get the customers back (Dennis 2007).

The last waste to be addressed is the *overprocessing* that occurs any time more work is done on a piece than what is required by the customer. When a product is processed beyond the standard that the consumer requires, all the additional processing costs are wasted. When the process is optimized, fewer resources are used to achieve the same level of customer satisfaction (Dennis 2007).

1.3.1.3 Committing to Lean Production

When all waste is eliminated from a manufacturing or service operation, the operation becomes much more efficient. Efficient production systems will comprise value-added steps throughout the operations (Baines et al. 2006). The key principle of Lean production is to cut out all the activities in a production line that do not add value to the product (Lee et al. 2008).

In the past, companies could use the formula: cost + profit margin = price. The Accounting Department would determine the costs and a profit margin typical for the type of industry would be used. In today's marketplace, the consumer will no longer be willing to pay the price without comparing the prices of similar products. The new profit equation will use the following formula: price (fixed) − cost = profit. This formula ensures that the only way that the profit can be increased is to reduce the cost. The cost reduction must be made in such a way that the company will stay profitable in the long term, maintenance must be performed, and the work team must be preserved. This ensures that the team increases their performance (Dennis 2007).

Lean transformation of a manufacturing or service enterprise is a process that takes a number of years from start to finish. The first step is to get commitment from the organization's executive management team. It is the executive management team that has the responsibilities for driving change and making the sometimes unpopular decisions that are required to keep the Lean initiative moving forward. It is important that the entire management team demonstrate their support and commitment with actions as well as words. The success of the Lean transformation relies on the total involvement of all employees (Nash and Poling 2007).

The next step is to conduct champion training for the key management champions. The value stream manager or champion that will be selected must know how to use the tools and techniques to identify the value-added activities along with the non-value-added activities within the operation. The value stream map of the current process with all of its inherent flaws included is used to drive the changes in subsequent iterations (Nash and Poling 2007). The value stream is defined from the customer's perspective. Once the value stream is identified, the seven wastes are eliminated and the process is organized to pull the work rather than push it (Seth and Gupta 2005).

The initial value stream map is used as a template to produce the envisioned future value stream map. It is important to realize that the part of the operation that is used must be small enough that the improvements are a doable list. When the list of improvements is too long or complicated at the implementation stage, many people may become despondent and frustrated about the possible success. When top management starts questioning the possibility of success, the Lean project may be jeopardized (Nash and Poling 2007).

1.3.1.4 Implementing a Lean Program

The important challenge when a manufacturing company wants to become Lean is the reduction of throughput time. The throughput time is reduced by the systematic removal of inventory. Reduction of inventory forces continuous changes in the manufacturing system design. In the Toyota system, inventory is controlled by the kanban system that limits the inventory that can be held between work cells. The system will also pass information upstream to inform those processes of what is required further along the production line (Black 2007).

The continued success of a manufacturing company depends upon the design of the manufacturing system. The manufacturing system is "a complex arrangement of physical elements characterized by measurable parameters" (Hunter and Black 2007, 130). The elements that are part of manufacturing systems include processes, tooling, material handling equipment, and people. When a system is altered, the behavior of the system will be changed (Hunter and Black 2007).

The most critical measurable parameter in the manufacturing system is time. When the system is manipulated to reduce the manufacturing interval systematically, some efficiency increases will occur. When the manufacturing time is reduced with the adoption of Lean production philosophies and methodologies, a higher level of efficiency will result with concomitant increases in international competitiveness (Hunter and Black 2007).

Lean production systems are not easy or cheap to implement and require determination by engineers and supervisors. The major subsystem of Lean production systems is cellular manufacturing. Lean production is flexible and can be adapted to different productive and competitive manufacturing environments. Cellular production concepts are simple and easy to adopt (Hunter and Black 2007).

One of the prime challenges when a company converts to Lean production is the change from final assembly to a mixed model to ensure that subassembly and component demand from suppliers are stable from day to day. This process is called leveling or smoothing of production. The subassembly lines are then converted into U-shaped cells to eliminate most or all of the conveyor lines. Each specific job shop is also converted to U-shaped cells that operate on a one-piece flow basis similar to the flow in final assembly. Kanban supply and demand systems are used to link subassembly and manufacturing cells to final assembly. This results in an integrated inventory and production control system (Black 2007).

1.3.1.5 Lean Production Implementation Steps

Top management must be committed to the opportunities and challenges of Lean manufacturing and they must communicate their commitment to all employees, especially to those in manufacturing. It is important to note that barriers to Lean implementation are frequently found in middle management since this is the group whose job functions will be impacted by Lean production systems (Black 2007).

The initial phase involves sharing the concepts of Lean production philosophy with everybody from the production workers to the company president. It is critical that the areas of cost, quality, technology standards, human resources, and information systems are in the loop. The total commitment of the leaders in the company is imperative. Success of the venture depends upon everybody's buy-in to the new way of operating. The old systems that were designed for mass production will no longer be useful and new systems will be developed to make financial decisions and to do the accounting of the venture (Black 2007). To reflect the financial performance of Lean production systems, Lean accounting is used to track value streams. The weekly value stream metrics that must be accounted include operational, capacity, and financial aspects of performance (Brosnahan 2008).

It is important to select the parameters that will be used to track the changes. The new paradigm that cost rather than price determines profits must be accepted by all. Everyone must also commit to eliminate waste in every part of the enterprise. All the production workers must be trained to recognize waste and to help find solutions to get rid of any waste. Operators must understand why the continuous change is vital for the success of the Lean processing system. The operators should also be empowered to design production cells, implement quality assurance methodology and machine tool maintenance, design control systems for production and inventory, reduce setup costs, and make continuous improvements later (Black 2007).

Once the Lean production process gets underway, it is important that the rewards that accompany the successful development of the Lean system are shared with the various teams. The gains caused by the efforts must be shared with all the people who contributed. In the case of middle management, the reward structure should support the system design (Black 2007). Performance-related pay is based on expectancy theory, where individuals are believed to exert more effort if they believe that the outcome of the exertion will have some value to them—in this case, more pay. But the causal simplicity posed by this basis does not account for the many other variables such as job characteristics and a perceived pay system that affects the outcomes of the performance-based-pay systems (Perry, Engbers, and Jun 2009).

1.3.1.6 Implementation Strategy

Many companies have implemented Lean manufacturing. The basic implementation strategies followed by the companies are outlined as ten steps by Black (2007).

During step one, manufacturing systems are leveled by developing mixed-model final assembly. The output from cells is balanced to match the requirements in final assembly. This is achieved by establishing daily demands and takt time for mixed-model final assembly, by balancing output from suppliers, by designing single-piece flow in subassemblies, and by sequencing subassemblies with order of assembly (Black 2007).

In step two, the manufacturing system is reconfigured into manufacturing and assembly cells (Black 2007). Manufacturing processes are more productive and efficiently managed when cellular manufacturing layout is used. Each cell comprises one or more machines that are able to perform more than one operation on the batch of material in the clustered group. When manual operations are used, multiskilled workers are required to complete all the required tasks. The machines in the cell can also be programmed, in which case the operations will be performed automatically. In this case, the worker has to perform only the setup for the particular batch. The cellular layout allows for a product layout that is used in flow lines and a functional layout used in job shop applications (Celano, Costa, and Fichera 2008). The manufacturing system design takes the product design, the functional requirements for system stability, and the needs of internal and external customers into account. Work is standardized for operators in each of the cells (Black 2007).

The setup time is reduced in step three to single digit times by changing methods and designs. The Lean paradigm acknowledges that the optimum lot size is one and the system design should be done to achieve single-minute exchange of dies. Operators must be led to develop one-touch setups in cells, and operators must be allowed to perform changeovers (Black 2007).

The quality assurance systems are integrated into the manufacturing system in step four, where the manufacturing process is designed to satisfy the design specifications of the product. Inspections are done to prevent defective products from leaving the manufacturing cell. The seven tools for quality control are used and the operators have the ability to stop the line when defects occur. The importance of quality is taught to everyone to ensure zero defects (Black 2007). The Taguchi model optimizes design settings while consideration is given to a large number of quality objectives at the same time. Response surface methodology is a statistical tool that is integrated with robust design techniques to find optimum operating conditions. Different characteristics in a product have conflicting objectives. When many characteristics are involved, the compromise between conflicting characteristics becomes very difficult to resolve while optimizing the process and the quality features. Kovach, Cho, and Antony (2008) developed an aggregate objective function that includes desirability ranges that are specified for each product function. Human judgment is supported in a way that will allow optimization of the production system and the quality features derived from consumers (Kovach et al. 2008).

Integrating preventative maintenance in the system is step five. Equipment is designed to be reliable and workers are trained to be reliable. Methodology must be in place to check people and the methods that they will use to check machines. The machines that are selected should be designed for total productive maintenance and operators should perform daily maintenance on them while solving any operational problems (Black 2007).

In step six, control mechanisms are put in place to control production in the cells regarding where, when, and how much material should be produced. The design of the system can use a process such as kanban to link cells and pull material to final assembly (Black 2007).

In step seven, the inventory control is integrated throughout the supply chain. Within the production system the work in progress between cells must be reduced to an optimized minimum. The gradual removal of inventory between cells will expose problem areas. Solving the problems will improve throughput time with limited inventory (Black 2007).

In step eight, the supply chain is addressed. Suppliers must be integrated into the manufacturing system with proper information flow to ensure that they become just-in-time manufacturers (Black 2007). Supply chain management is the collaborative coordination between companies to leverage strategic positioning for the ultimate improvement of operating efficiency. To ensure optimal efficiency, supply chain operations require operational management processes that cross functional boundaries within organizations and link suppliers, partners, and consumers across organizational boundaries (Bowersox, Gloss, and Cooper 2007). It is imperative that suppliers are successful since Lean depends upon sole-source suppliers with long-term good relationships (Black 2007).

The next step, step nine, addresses the issue of allowing people decision-making capability for quality and quantity of production. This ability to be autonomous comes after functional integration has occurred. At the same time, automation of integrated pull systems is put in place. For this, robots, computers, and task automation are used (Black 2007). Success is not represented by incorporation of information technology unless it is accompanied by Lean engineering and acknowledgment of the benefits that can be gained by being more responsive to the needs of the consumer (Accacia et al. 1995).

In the final step, cell designs are improved to take them from interim cells that use single-cycle automatics to Lean cells that are designed for single-piece flow. New products are designed concurrently with customers to ensure manufacturability and customer support. Lean manufacturing is designed and implemented with Lean machine tools (Black 2007).

1.3.2 Quantifying the Value of Lean Manufacturing

Changing from a well-known production system to a Lean production system is a difficult decision and a long and expensive process. There are substantial differences between traditional and Lean manufacturing systems regarding the management of employees, production plant layout, and inventory and information systems, as well as production scheduling methods. The enormous undertaking to implement a Lean manufacturing process is normally done by guessing what the potential benefits may be. Many of the benefits, such as decreased lead time for customers, reduced inventories for manufacturers, and improved information management, are valuable but difficult to quantify (Melton 2005).

The decision to change a company from a traditional to a Lean enterprise is frequently based on a combination of conviction that the Lean manufacturing philosophy will be good, on reported experiences of others who adopted these principles, and on assumed best-guess anticipated benefits. Few management teams will be swayed to adopt Lean concepts on these premises (Detty and Yingling 2000).

It is possible to use simulation to quantify expected performance improvements gained by applying Lean manufacturing shop floor principles, just-in-time inventory systems, continuous quality assurance, and level production scheduling. The simulation can also be expanded to include warehousing and work-in-process inventory, material movement requirements, the value of production control and scheduling systems, uniform supply from material suppliers, and system responsiveness to market demand (Detty and Yingling 2000).

When compared to the system that was in place, the Lean production system yielded many beneficial changes in the production system. The average time that a single component spends in the system was reduced 55 percent. Changeover times were reduced from 11 to 3 minutes. Inventories in the warehouse were reduced by 70 percent and the assembly cell inventories were reduced by 75 percent. The finished goods inventory was reduced by 10 percent. Seven forklift trucks were not used and the seven forklift drivers and nine employees from preassembly and kitting could be reassigned to other activities. The sixteen employees constituted a large savings benefit (Detty and Yingling 2000).

1.3.2.1 Lean Accounting

The primary goal of Lean manufacturing is to reduce waste and provide increased customer value. All the human resources strive to increase output, productivity, and quality while inventory, defects, floor space utilization, cycle time, and rework are reduced. When a company starts operating Lean production, the financial statements for the company frequently see a drop in profits when deferred labor and overhead move from the inventory account on the balance sheet to the expense section of the income statement. The true value of Lean is frequently hidden in the financial statements. The problem in accounting is to show that the true financial performance of a company that embarked on Lean production has been solved by moving away from standard cost accounting to a new accounting format that tracks Lean costs and benefits more accurately (Kroll 2004).

Accountants that have worked with Lean companies agree that many standard cost accounting practices do not track the performance of the operation accurately. Traditional cost accounting was developed to report the value of a company to outsiders. The purpose of the report was not to help managers to enhance the operation by removing waste. Since traditional accounting assumptions contradict Lean manufacturing, Lean companies are implementing Lean accounting to get a better understanding of the value of the enterprise (Kroll 2004).

In regular cost accounting, nonfinancial measures such as reduced lead times, reduced scrap rates, and on-time deliveries are not captured on GAAP (generally accepted accounting principles) financial statements, while they show significant improvements within the Lean enterprise. The net income may decline for the short term when the company launches Lean manufacturing. During the initial stages, existing inventory must be used and deferred labor and overhead move from the asset side of the balance sheet to the expense part of the income statement (Kroll 2004).

1.3.3 The Skinny on Lean Implementation

When a company embarks on Lean manufacturing, the top management will have the following goals in mind. A production system based on total quality is able to produce products that are perfect the first time. This quest for zero defects is inherently part of the quest for higher productivity. It is also important that the source of product failure is identified. The workers themselves are a part of the problem solution and this empowers the workers, leading to enhanced teamwork, motivation, and productivity. Another general goal will be the minimization of waste by eliminating all activities that do not create value. At the same time scarce resources will be used efficiently to enhance the return on investment. The improvements gained with Lean are assumed to be continuous and the Lean process will continuously evolve to reduce costs, improve quality, and increase productivity. Another goal is to deliver

- Flexibility in producing different mixes or greater diversity of products quickly, without sacrificing efficiency at lower volumes of production, through rapid setup and manufacturing at small lot sizes
- Long-term relationships between suppliers and primary producers (assemblers, system integrators) through collaborative risk-sharing, cost-sharing, and information-sharing arrangements

Lean as a manufacturing system is a concept rather than a process. There is no recipe or a cookie-cutter list of steps that can be applied to a company to move it from traditional operations to Lean. The paradigm shift that will make things happen is an acceptance of the Lean philosophy by everyone in the company. Acceptance of the Lean philosophy depends upon the way in which the Lean philosophy is introduced and sold to everyone in the company. It is when everyone realizes the potential benefits of the new way of doing that the change to Lean production will begin. It is also a process that takes a long time to get to the stage where it shows any benefits. Good change management is the key to the success in transition from a traditional to a Lean operation.

Within the boundaries of corporate culture, there are leaders and followers. Although followers will frequently follow one leader, that leader operates within the constraints set by peers and superiors. Organizations that embark on transition to Lean operations are doing so to enhance performance of employees to ultimately benefit the company and the stakeholders in the company. Leadership as a change agent is essential to drive acceptance of the Lean philosophy forward within the framework of project management, education, and organizational intelligence within the constraints of corporate social responsibility

In the case of Lean thinking, the leader will become a champion for the transition. Many of the change agents will lead the different groups within the organization to the unified goal of Lean manufacturing. These change agents will be led by top management that needs to buy into the Lean philosophy. The change agents as leaders will lead other people as a group from the present traditional manufacturing system to a situation where everyone accepts that Lean is a way to move the company forward with ultimate benefits for all.

Managers identified inspiration, strategic thinking, forward-looking, honesty, fair-mindedness, courage, supportiveness, and knowledge as the key characteristics of leadership (Hewison 2004). To promote acceptance of the Lean transition, it is vital to make a clear distinction between managers and leaders. Although all managers must lead and all leaders must be competent managers, the responsibilities of being the change agents to drive the buy-in to Lean are outside the realm of just managing a group of people. The relational leadership style is required to be effective in this case. Relational leaders have vision and inspiration but act more as enablers. They can lead as easily as they can follow; they are team players and empower the people they work with. Relational leaders produce stability in the wake of changes and their leadership style is frequently invisible to most observers (Hewison 2004).

The benefits of the change to Lean operations should be applied to organizational metrics. The metrics normally include reduction in failure rates, enhanced delivery and customer service, increased reliability of deliverables, the deployment of resources to the greatest advantage to the company, and increased return on investment. There are also inherent benefits on the human side of the company. Lean philosophy and Lean thinking will affect the problem-solving potential of the people that work within the Lean paradigm. People who have undergone the change to Lean operations develop a new appreciation of the benefits that accompany the acceptance of different ways to perform their tasks. The best way to demonstrate the financial impact of Lean manufacturing on the profitability of the enterprise is to use Lean accounting. Lean accounting measures and evaluates a company's performance by measuring value stream metrics. It is an acceptance that change is good and that everyone has the potential to suggest a new way of doing a job that will benefit the company.

Whatever the process to move a company from the traditional style of manufacturing to Lean manufacturing is, it is one way to enable US-based manufacturing companies to compete on a global scale. The new way of creating value through Lean manufacturing is revolutionizing manufacturing and the leaders of the companies that adopt this new way to do business will have to continue to lead the company to become more Lean. Going Lean is a journey of continuous change with concomitant financial, cultural, and organizational benefits.

The challenge of Lean is to keep on moving the changes forward without the luxury of sitting back and being proud that you have arrived. Change is not easy to accept, and it is not ever a comfortable place to be. But embarking on Lean will take the whole enterprise into the unknown.

1.3.3.1 Strategy to Move Lean Thinking Forward

5S is a way of doing that was developed in Japan. The "just in time" (JIT) methodology is based on 5S. JIT is used to reduce waste and optimize productivity by keeping the workplace ordered and using visual cues to ensure consistent operational results. Implementing of 5S methodology is simple:

1. Sort (*seiri*)
2. Set in order (*seiton*)
3. Shine (*seiso*)
4. Standardize (*seiketsu*)
5. Sustain (*shitsuke*)

Using the five simple acts to organize, clean up, develop, and sustain the production environment will result in time economies and enhanced production efficiency. The method is used to encourage workers to improve their working environments with concomitant reduction in waste, unplanned downtime, and in-process inventory (EPA 2011).

Implementing 5S methodology will reduce the space required for production and is a way to increase production in established space. The tools and materials will be organized in labeled and color-coded storage locations. Everything required to perform a specific production operation will be combined in a kit that can be used when required (EPA 2011).

Sort is the action of reducing items in the workplace by elimination of all the unnecessary items. Items that are not required for the operations in the area are red tagged. Once an item is red tagged, it is moved to a holding area where the item is disposed or reassigned to another area. After sorting, valuable floor space and tool space are retrieved and broken tools, scrap, and excess raw material are removed (EPA 2011). It is important to realize that essential tools may have to be duplicated in each area since finding them somewhere else may waste time. So this Lean thinking may only work when additional tools are acquired.

To *set in order* means that efficient and effective storage methods are created to arrange items for easy use and to label them for easy location and replacing. This step can only be implemented after sort has occurred. In this step it is normal to allocate areas for everything by painting floors, labeling tools and equipment, and designating a storage location for each tool. Work areas will be outlined and modular shelving and cabinets will be installed (EPA 2011).

The *shine* step relates to the organization and thorough cleaning of the work areas. The cleanup is not a onetime activity, but rather an activity that is done daily until the workplace is clean. The clean area is essential for workers to identify any problems such as leaks, abnormal vibrations, breakages, and misalignments that would eventually lead to equipment failure and production losses. The shine targets—assignments, methods, and tools—are normally assigned prior to the establishment of the shine pillar (EPA 2011).

Once sort, set in order, and shine are implemented, the next pillar is standardization of best practices in the work area. *Standardize* is the way to maintain the first three pillars by creating consistent methodology to achieve each task or procedure. This involves job cycle charts, visual cues such as signs and display scoreboards, scheduling of 5S periods, and check lists. The second part of standardize is prevention of accumulation of surplus items, preventative maintenance, and constant cleaning (EPA 2011).

Finally, the system should be sustained. To *sustain* the use of the correct procedures over time is frequently the most difficult S to maintain. The change of a comfortable way of working to one that requires constant alertness to the Lean goals is difficult. To ensure that the new status quo is sustained requires constant assurance that there is no reversion to the good old times. The five pillars are mutually inclusive; if one of them is not maintained, the system fails as a whole. Tools to sustain 5S include signs and posters, newsletters, pocket manuals, team and management check-ins, performance reviews, and department tours. It is essential that the 5S system be reinforced by messages in all appropriate formats until 5S become the accepted and standard operating procedure (EPA 2011).

Questions

1. List your actions from the time you get up in the morning until you finally leave home.
   a. How could you shorten this period?
   b. How much time could be saved?
   c. How much time could be saved in a year with 200 workdays?
2. When you are looking for something in your toolbox or handbag, what is the most frustrating element regarding the storage of the items?
3. Choose any service place such as a hairdresser or nail salon. Go there as a customer and observe all the possible areas of waste. Describe how you would implement a Lean system there.
4. How would you start a Lean system for the mail service of the Post Office?

References

Accacia, G. M., Callegari, M., Michelini, R. C., Milanesio, R., Molfino, R. M., and Rossi, A. (1995). Pilot CIM implementation for Lean engineering experimentation. *Computer Integrated Manufacturing Systems* 8 (3): 185–192.

Baines, T., Lightfoot, H., Williams, G. M., and Greenough, R. (2006). State-of-the-art in Lean design engineering; a literature review on white collar Lean. *Journal of Engineering Manufacture* 220:1539–1547.

Black, J. T. (2007). Design rules for implementing the Toyota production system. *International Journal of Production Research* 45 (16): 3639–3664.

Bowersox, D. J., Gloss, D. J., and Cooper, M. B. (2007). Supply chain logistics management (6th ed.). McGraw–Hill Irwin, Boston MA.

Brosnahan, J. P. (2008). Unleash the power of Lean accounting. *Journal of Accountancy* 206 (1): 60–66.

Celano, G., Costa, A., and Fichera, S. (2008). Scheduling of unrelated parallel manufacturing cells with limited human resources. *International Journal of Production Research* 46 (2): 405–427.

Chongvilaivan, A., Hur, J., and Riyanto, Y. E. (2009). Outsourcing types, relative wages, and the demand for skilled workers: New evidence from U.S. manufacturing. *Economic Inquiry* 47 (1): 18–33.

Conti, J. P. (2009). Hidden dragon. *Engineering and Technology* 4 (2): 64–66.

Dennis, P. (2007). *Lean production simplified* (2nd ed.). Productivity Press, New York.

Detty, R. B., and Yingling, J. C. (2000). Quantifying benefits of conversion to Lean manufacturing with discrete event simulation: a case study. *International Journal of Production Research* 38 (2): 429–445.

EPA. (2011). *Lean thinking and methods*. Retrieved from http://www.epa.gov/lean/environment/methods/fives.htm

Flores, J. (2008). Employee benefit plan review. *Benefits Quarterly* 24 (4): 50.

Ford. (2009). Innovator, industrialist, outdoorsman: Henry Ford started it all. Retrieved on November 28, 2009, from http://www.ford.com/about-ford/heritage/people/henryford/650-henry-ford

Greasley, D., and Oxley, L. (1996). Explaining the United States' industrial growth, 1860–1991: Endogenous versus exogenous models [electronic version]. *Bulletin of Economic Research* 48 (1): 65–82.

Hewison, R. (2004). The crises of cultural leadership in Britain. *International Journal of Cultural Policy* 10:157–166.

Hunter, S. L., and Black, J. T. (2007). Lean remanufacturing: A cellular case study. *Journal of Advanced Manufacturing Systems* 6 (2): 129–144.

Kovach, J., Cho, B. R., and Antony, J. (2008). Development of an experiment-based robust design paradigm for multiple quality characteristics using physical programming. *International Journal of Advanced Manufacturing Technology* 35:1100–1112.

Krajewski, L. J., Ritzman, L. P., and Malhotra, M. K. (2007). Operations management (8th ed.). Prentice Hall, Upper Saddle River, NJ.

Kroll, K. M. (2004). The lowdown on Lean accounting. *Journal of Accountancy* 198 (1): 69–76.

Lee, Q. (2008). Lean manufacturing essentials. *Management Services* 52(2): 46–47.

Lee, S. M., Olson, D. L., Lee, S-H, Hwang, T., and Shin, M. S. (2008). Entrepreneurial applications of the Lean approach to service industries. *The Service Industries Journal* 28 (7): 973–987.

McDonald, K. M. (2009). Do auto recalls benefit the public? *Regulation* 32 (2): 12–17.

Mehta, M. (2009). A + E = LEAN. *Industrial Engineer* 41 (6): 28–33.

Melton, T. (2005). The benefits of Lean manufacturing. What Lean thinking has to offer the process industries. *Chemical Engineering Research and Design* 86 (6): 662–673.

Nash, M., and Poling, S. R. (2007). Strategic management of Lean. *Quality* 46 (4): 46–49.

Perry, J. L., Engbers, T. A., and Jun, S. Y. (2009). Back to the future? Performance-related pay, empirical research, and the perils of persistence. *Public Administration Review* 69 (1): 39–51.

Schroeder, R. G. (2008). Operations management: Contemporary concepts and cases (4th ed.). McGraw–Hill Irwin, Boston, MA.

Serrano, I., Ochoa, C., and De Castro, R. (2008). Evaluation of value stream mapping in manufacturing system redesign. *International Journal of Production Research* 46 (16): 4409–4430.

Seth, D., and Gupta, V. (2005). Application of value stream mapping for Lean operations and cycle time reduction: An Indian case study. *Production Planning and Control* 16 (1): 44–59.

Shang, W., and Hooker, N. H. (2005). Improving recall crisis management: Should retailer information be disclosed? *Journal of Public Affairs* 5 (3): 329–341.

Shingo, S. (1989). *A study of the Toyota production system from an industrial engineering point of view.* Productivity Press, New York.

Spear, S. J. (2004). Learning to lead at Toyota. *Harvard Business Review* 82 (5): 78–86.

US Department of Labor. (2009). Wage and hour division. Retrieved on December 12, 2009, from http://www.dol.gov/whd/flsa/index.htm

Womack, J. P. (2007). Moving beyond the tool age [electronic copy]. *Manufacturing Engineer* 86 (1): 4–5.

Wood, N. (2004). Lean thinking: What it is and what it isn't. *Management Services* 48 (2): 8–10.

Wu, S., Wang, S., Blos, M. F., and Wee, H. M. (2007). Can the big 3 overtake Toyota?—A study based on the theory of constraints [electronic version]. *Journal of Advanced Manufacturing Systems* 6 (2): 145–157.

Chapter 2

Measurements and Numbers

Many people have an aversion to anything that involves numbers. It seems that written descriptive problems are even more of a challenge. In life, many challenges start as concepts that we solve by using calculations. If you want to win when playing poker, it may help if you know the statistical probability that you have a winning hand. It is easy to know what the chances are of drawing a specific card when playing with a single deck. Mental calculation of the probability that someone else may have a better hand than your own will go a long way to ensure success in the outcome. When we work around the home, we frequently need to calculate surface area before painting or carpeting. Calculations are easy once you realize that they are based on a language that requires a good understanding of the terminology.

Scientists communicate through the common language of numbers and it is important to quantify results as numbers. To make sense of values, it is important to describe what the value addresses accurately. Any number can have different meanings when the context of where the number is used is taken into account. The number 5 can be used on a door, a coin, a ticket, a stamp, an invoice, etc. To ascribe a meaning to the number, we have to look at the situation and application since it represents differently applied concepts such as counting and cardinality,* labeling, measure, date, and time (Rogers 2008).

---

* Cardinality in a linguistic sense refers to the words for numbers—for example, one, two, three—while in its mathematical use it is a generalization of the natural numbers used to measure the cardinality or size of sets, such as three numbers in the set.

The key to using numbers in a logical way to solve problems is embedded in your proficiency with number sense. According to Faulkner (2009), number sense refers to

■ Fluency in estimating and judging magnitude
■ Ability to recognize unreasonable results
■ Flexibility while using mental computing
■ Ability to move among different representations and to use the most appropriate representations

To increase your proficiency, you will have to exercise your number skills. I find that playing any kind of number game increases my skills with numbers. Sudoku is a good place to start and it is a great way to keep your brain in good shape.

Measurements of some sorts have been used to describe quantities and sizes of artifacts from the earliest times. If one considers the size of the fish that got away, it becomes clear just how important it is to have a clear definition of size.

It is overwhelming to look at the Eiffel Tower while considering how many pieces of metal had to be precut and predrilled to make up that gigantic three-dimensional masterpiece. Even more important, the pieces had to fit in a specific spot. Doing the calculations of the forces involved and combining them with material strength and rigidity would be a huge undertaking with the aid of computers. How about doing it all on paper?

2.1 History

Measurements relate to the world as we observe it and a standard needs to be chosen. In England, the inch was defined as the length of three grains of barley. Since nature is not very consistent in the length of barley, the length of an inch changed from year to year. A grain was the weight of a grain of pepper and so forth. There is a rather amusing tale regarding the US Standard railroad gauge* (the distance between the rails) of 4 feet, 8½ inches. Stephenson chose the exceedingly odd gauge, used in many countries, since it was about the standard axle length for wagons. The people who built the tramways used the same jigs and tools that they had used for building wagons.

The ruts in the unpaved roads dictated the wheel spacing on wagons. If they used any other spacing, the wheels would break since they did not fit into the ruts made by countless other wagons and carriages. Roman war chariots made the initial ruts, which everyone else had to match for fear of destroying the wheels on their wagons. Since the chariots were made for or by Imperial Rome, they were all

---

* The URL for this information is http://www.straightdope.com/columns/000218.html

the same. The railroad gauge is thus derived from the specifications for an Imperial Roman army war chariot. The Romans used this particular axle length to accommodate the back ends of two warhorses. The space shuttle had two booster rockets of 12.17 feet in diameter attached to the sides of the main fuel tank. These solid rocket boosters were made by Thiokol at a factory in Utah. The engineers who designed the boosters had to ensure that the boosters would fit through a tunnel. Railway tunnels are designed based on the width and height of the rail cars, and that is based on the gauge of the track. In a strange way, what was one of the world's most advanced transportation systems had a major component that was indirectly determined by the size of the rumps of two Roman warhorses.

2.2 Standards

In medieval times, measures were standardized in a market town. The standards could also vary from one town to another. Standardization started with a royal decree that could vary from time to time. With the start of international trade, a bolt of cloth or a sack of wool was taxed based on size, and standard measures became more important. After the French revolution, the French people associated the units of measurement with the oppression of the *ancien régime* and demanded new units. The introduction of the meter was not very popular in France and there were periods when France reverted to the old measures. The scientific community acknowledged the benefits of a decimal system and the adoption spread through Europe until it was officially adopted as an international measurement unit by the Meter Convention of 1875.

A problem with the establishment of a measure is that someone has to choose a standard and that others will use it. The meter was chosen to be 1×10^{-7} of the distance from the North Pole to the equator along a longitudinal line running near Dunkirk. The units for volume and mass were derived from the meter. One cubic decimeter was chosen as the liter and the mass of water at 4°C contained in one cubic centimeter was chosen as a gram. These original measures were replaced by standards that are much more accurate.

Since people, particularly engineers and scientists, need to communicate through numbers, it is important to define what numbers *mean*. In the same trend, to make sense of quantity one must define accurately what measurements mean. This book uses the SI system since the preferred metric units are multiples of 1000, making conversion a simple matter of changing exponents by adding or subtracting 3:

$$1,000,000 \text{ mm} = 10^6 \text{ mm} = 10^{6-3} \text{ m} = 10^3 \text{ m} = 10^{3-3} \text{ km} = 10^0 \text{ km} = 1 \text{ km}$$

Some conversion tables are incorporated in Appendix 2. Please note the different notations that can be used. The solidus notation (i.e., m/s or g/kg or m/s/s) can be written in negative index notation as m s^{-1}, g kg^{-1}, and m s^{-2}. In the SI system, the negative index notation is preferred.

2.3 Dimensions

Dimension refers to any one of the three physical or spatial properties of length, area, and volume. In geometry, a point is said to have zero dimension. A figure having only length, such as a line, has one dimension. A plane or surface has two dimensions, and a figure having volume has three dimensions. The fourth dimension is often said to be time, as in the theory of general relativity. Higher dimensions can be dealt with mathematically but cannot be represented visually.

There is some very interesting work done in physics that is turning my mind:

> We study higher dimensional scalar–tensor theory with matter. The dimension is ten and it is anisotropic in the sense that the dimension has a product structure both with four-dimensional space-time which plays a role of our universe and the extra six-dimensional space which should be too small to be observed. The matter is also taken as an anisotropic. The four dimensions are expanding while the extra six dimensions are contracting for special choice of matter (Lee 2011, 20).

Dimensions have more than one definition. In mathematics, the geometric dimension of a set is defined by a natural number. For a single point, the dimension is zero, while a straight line has a dimension of one. A plane will have two dimensions and a volume will have three dimensions (Debnath 2006). The dimension of an object is defined as the minimum number of coordinates needed to specify any point within it. To find any point on a straight line, we need only one coordinate. Where is 5 on a ruler?

A second definition for dimension is any measurable extent such as time, distance, mass, volume, or force. Remember that these dimensions are a combination of the basic physical dimensions of mass, length, time, electric charge, and temperature. Speed is measured in the dimensions of length per unit time and may be measured in meters per second. Units are designations of the amounts of a specified dimension such as second, meter, liter, and newton. The units and preferred symbols for the dimensions are given in Table 2.1.

Some conversion tables are included in Appendix 1.

2.4 Units

Units are designations of the amounts of a specified dimension such as second, meter, liter, and newton. It is important that one keeps a sense of sanity within the very confusing system of units (numbers) with different dimensions (measured extent) associated with them. The easiest way to do this is with dimensional analysis.

Table 2.1   Dimensions and Preferred Units Used in SI System

Dimension	Unit	Preferred Symbol
Mass	Kilogram	kg
Length	Meter	m
Time	Second	s
Absolute temperature	Kelvin	K
Temperature	Degrees Celsius	°C = K − 273
Electric current	Ampere	A
Luminous intensity	Candela	cd
Amount of a substance	Mole	mol

2.4.1 Base Units

There are seven base units in the measurement system. The first four—mass, length, time, and temperature—are used by everyone while the last three—electric current, luminous intensity, and amount of a substance—are used in technical and scientific environments.

Each of the base units is defined by a standard that gives an exact value for the unit. There are also two supplementary units for measuring angles.

2.4.2 Definitions

Most of the dimensions were selected in some arbitrary fashion. The Fahrenheit scale was chosen such that the temperature for a normal human would be 100°F; eventually the mistake was discovered and we now have an average normal temperature of 98.6°F. There are many other legends about the choice of the standard foot, inch, drams, crocks, grains, and bushels. When the metric system was defined, some standards were chosen in an arbitrary way that might have been founded in a scientific misconception. In most cases, the older standards that relied on artifacts that are kept somewhere have been replaced by a more scientific definition. As our knowledge of science expands, new definitions will be formulated for standards.

Time[*]: this is a definite portion of duration—for example, the hour, minute, or second. The old system used a mean solar second as equal to 1/86,400 of a mean solar day. Now a second is defined as the time it takes for 9,192,631,770 electromagnetic radiation waves to be emitted from ^{133}Cs.

---

[*] For a good treatise on time, Paul-Henning Kamp has written thought-provoking papers regarding time and computers.

Mass[*]: this is a measure of the number of molecules present in a body. The mass in this case will be the inertial mass that represents a measure of a body's resistance to changing its state of motion when a force is applied (Fiorentin 2009). The mass of a body is thus independent of gravity; it is the same anywhere in the universe. The weight of a body is measured in newtons and is the force of gravity acting on the body. The weight of a body will change when there is a change in gravity. The gravitational field on Earth is 10 N/kg, so the weight of 1 kg is 10 N.

A kilogram is defined as the mass of a block of platinum-iridium kept at St. Sèvres in France.

Length: The meter used to be defined as the length of a platinum-iridium bar kept at St. Sèvres in France or the wavelength of orange light emitted by a discharge lamp containing pure krypton at a specific temperature, which is 6.058×10^{-7} m. The new definition for a meter is the length of the path traveled by light in vacuum during a time interval of 1/299792458 of a second.

Temperature: this is the degree of hotness or thermodynamic temperature of a body. The kelvin is the fraction 1/273.16 of the thermodynamic temperature of the triple point of water. The triple point occurs at a pressure of 760 mm Hg and a temperature such that liquid water, solid ice, and water vapor can coexist in a stable equilibrium. This is at 273.16 K. Remember that, in degrees Celsius, that will be at 0.01°C.

The two fixed reference points for the Celsius scale are the melting point and boiling point of pure water at one atmosphere or 760 mm Hg pressure. These temperatures in Celsius are 0°C and 100°C. Another frequently used reference point is the low temperature where all molecular motion will stop, 0 K or –273.15°C.

Electric current: Ampere is a measure of flow of electrons. The ampere is that constant current that, if maintained in two straight parallel conductors of infinite length and negligible circular cross section, and placed 1 m apart in vacuum, would produce between these conductors a force equal to 2×10^{-7} N per meter of length.

Luminous intensity: Candela is the luminous intensity, in a given direction, of a source that emits monochromatic radiation of frequency 540×10^{12} Hz and that has a radiant intensity in that direction of 1/683 watt per steradian.

Amount of a substance: One mole is the amount of a substance of a system that contains as many elementary entities as there are atoms in 12×10^{-3} kg of ^{12}C. When the mole is used, the elementary entities must be specified and may be atoms, molecules, ions, electrons, other particles, or specified groups of such particles. This corresponds to a value of $6.02214179(30) \times 10^{23}$ elementary entities of the substance. In chemistry we frequently refer to mol per liter, or in preferred SI as mol dm^{-3}.

---

[*] Mass is not very well defined and even Einstein tried to define mass in 1905. Look at the paper by Fiorentin for a good description of the work done to define mass.

2.4.3 Supplementary Units

Circular measure: This is mathematically convenient to measure plane angles in dimensionless units. The *natural* measurement is length of circular arc divided by the radius of the circle. Since circumference of a circle with radius R is given by $2\pi R$, $360° = 2\pi$ radians.

Steradian: measurement of solid angles, equal to the angle subtended at the center of a sphere of unit radius by unit area on the surface. A steradian is dimensionless since 1 sr = $m^2 \bullet m^{-2}$ = 1. The symbol sr is used when appropriate to distinguish between dimensionless quantities of different natures.

2.4.4 Derived Units

The clear definitions of the basic units allow for standardization of the dimensions. Mathematical combinations of standard base units are used to formulate derived units.

Area: product of two lengths describing the area in question (m^2). For rectangular units it is given by width × length and for circles it is πr^2. Land is frequently measured in hectares with 10,000 m^2 equal to 1 hectare equal to about 2.5 acres. Large land areas are measured in square kilometers.

Volume: product of three lengths (m^3) describing the space in question:

$$1\ m^3 = 1\ m \times 1\ m \times 1\ m = 10^3\ mm \times 10^3\ mm \times 10^3\ mm \times 10^9\ mm^3 = 10^6\ cm^3 = 10^3\ dm^3 = 10^3\ liters$$

Density: the mass of a substance divided by its volume:

$$\rho = \frac{kg}{m^3} = kg\ m^{-3}$$

Remember to change the sign of the exponent when a unit is inverted (moved from below the line to above the line).

Velocity: distance traveled in a specific direction (m) divided by the time taken to cover the distance (s):

$$\frac{m}{s} = V \quad \text{or} \quad ms^{-1} = V$$

Velocity is a vector because it is directional. Other vector quantities include force, weight, and momentum. Kinetic energy and speed have no direction associated with them and are scalar quantities. Other scalar quantities include mass, temperature, and energy.

Flow: this is cubic meters per second (m^3s^{-1}) are also referred to as cumec.

Momentum: linear momentum is the product of mass and velocity ($kg\ ms^{-1}$ or Ns).

Acceleration: the rate of change of velocity of a body is expressed in SI as ms^{-2}.

Gravitational constant: $g = 9.81\ ms^{-2}$.

Force: this is the product of mass (m) and acceleration (a):

$$F = ma$$

F, the resultant force on the body, is measured in newtons, *m* is the mass in kilograms, and *a* is the ultimate acceleration in meters per second per second ($N = 1\ kg\ ms^{-2}$).

If a body of mass *m* starts from rest and reaches a velocity *v* in *t* seconds as a result of force *F* acting on it, then the acceleration is *v/t* and

$$F = ma = \frac{mv}{t} \text{ the rate of change in momentum}$$

Pressure: this is the force applied on a specific area ($N\ m^{-2}$), also known as pascal (Pa). This is a very small unit and the practical units are kilopascals or bar. One bar is equal to approximately one atmosphere:

$$1\ bar = 10^5\ Pa\ (Nm^{-2}) = 0.1\ MPa \cong 1\ atmosphere$$

Frequency: this is cycles per second = hertz (Hz); $10^6\ Hz = 1\ MHz$

2.5 Dimensional Analysis

When you work with numbers, it is important to remember that the unit (number) is associated with a dimension (something that is measured). If a bag of fruit contains two apples and three peaches, nobody will try to add the apples and peaches together and call them five bananas. In the same way, we can only do calculations with numbers that belong together.

Example 2.1

Convert 25,789 g to kilograms.

$$25789\ g = \left(\frac{25789\ g}{1} \right)\left(\frac{1\ kg}{1000\ g} \right) = 25.789\ kg$$

When we work with numbers that have different dimensions, they cannot be manipulated unless the dimensions are part of the manipulation. From basic definitions, we learn that density is mass/volume, so

$$\text{density} = \frac{\text{kg}}{\text{m}^3} \text{ or in the more usual terms, density} = \frac{\text{g}}{\text{cm}^3}$$

Example 2.2

It is most important to give the density base since 1000 g = 1 kg and $(100)^3$ cm^3 = 1 m^3. When we convert a density of 1500 kg/m^3 to grams per cubic centimeter or the frequently used grams per milliliter, we do it as follows:

$$\frac{1500 \text{ kg}}{\text{m}^3} = \frac{1500 \times 1000 \text{ g}}{100 \text{ cm} \times 100 \text{ cm} \times 100 \text{ cm}} = \frac{1,500,000 \text{ g}}{1,000,000 \text{ cm}^3} = \frac{1.5 \text{ g}}{\text{cm}^3} = 1.5 \text{ g/ml}$$

Example 2.3

Convert 20 kPa to kilograms per meter per second squared.

$$20 \text{ kPa} = (20 \text{ kPa})\left(\frac{1000 \text{ Pa}}{1 \text{ kPa}}\right) = 20000 \text{ Pa} = (20000 \text{ Pa})\left(\frac{1 \text{ Nm}^{-2}}{1 \text{ Pa}}\right)$$

$$20 \text{ kPa} = (20000 \text{ Nm}^{-2})\left(\frac{1 \text{ kg ms}^{-2}}{1 \text{ N}}\right) = 20000 \text{ kg m}^{-1}\text{s}^{-2} = 20000 \text{ kg/ms}^2$$

Example 2.4

Calculate the mass of aluminum plate with the following dimensions: length = 1.5 m, width = 15 cm, and thickness = 1.8 mm. The density of the aluminum = 2.7 g/cm^3.

$$\text{mass} = \text{volume} \times \text{density}$$

$$\text{mass(g)} = \frac{150 \text{ cm} \times 15 \text{ cm} \times 0.18 \text{ cm}}{1} \times \frac{2.7 \text{ g}}{1 \text{ cm}^3}$$

$$\text{mass(g)} = \frac{405 \text{ cm}^3}{1} \times \frac{2.7 \text{ g}}{1 \text{ cm}^3} = 1093.5 \text{ g}$$

The dimensions can be canceled just as we do in calculations. The cubic centimeters below the division line will cancel out the same term above the division line. The same calculation done in SI preferred units would be as follows:

$$\text{mass(kg)} = \text{volume(m}^3) \times \text{density} \left(\frac{\text{kg}}{\text{m}^3} \right)$$

$$\text{mass(kg)} = \frac{1.5 \text{ m} \times 0.15 \text{ m} \times 0.0018 \text{ m}}{1} \times \frac{0.0027 \text{ kg}}{0.0000001 \text{ m}^3}$$

$$\text{mass(kg)} = \frac{0.000405 \text{ m}^3}{1} \times \frac{2700 \text{ kg}}{\text{m}^3} = 1.0935 \text{ kg}$$

2.6 Units for Mechanical Systems

Work: mechanical work (J) is force × distance moved in the direction of the force. Work is measured in joules (J), where 1 J of work is done when a force of 1 N moves its point of application through 1 m.

$$J = N \times m = kg \ m^2 \ s^{-2}$$

Example 2.5

The work to lift 1 kg of material a height of 3 m is

$$1 \text{ kg} \times 3 \text{ m} \times 9.81 \text{ ms}^{-2} = 29.43 \text{ kg m}^2 \text{ s}^{-2} = 29.43 \text{ J}$$

Power: this is the rate of work done on an object. One watt (W) is the power expended when 1 J of work is performed in 1 s (W = J s^{-1}).

Throughout the ages, the power unit on farms was a horse or an ox. When mechanical power units were introduced, it was natural that the power of the steam contraption was compared to that of a horse. Very few people know the kilowatt output of the engine in their automobiles, but most will be able to tell you what it is in horsepower (hp). It is rather strange that we frequently use different measuring systems within one object. The size dimensions of the automobile are given in inches and cubic feet. The engine displacement is given in *cc*, which is metric and a way to denote cubic centimeters. One of the biggest wastes of time and effort is in the continual conversion of units. One horsepower (hp) is equivalent to 550 ft lb$_f$ s^{-1}. The conversion of horsepower to joule is as follows:

$$1 \text{ hp} \Leftrightarrow 745.7 \text{ W} \Leftrightarrow 745.7 \text{ J s}^{-1}$$

Example 2.6

In conversion problems, dimensions and units become even more confusing. The conversion factor when one converts 1 horsepower (hp) to watt (W) is given as 1 hp = 745.7 W.

The calculation with dimensional analysis will look like this:

$$1 \text{ hp} = 550 \text{ ft lb}_f/s$$

$$\text{ft lb}_f = \text{lb}_m \times \frac{\text{ft}}{s^2} \times g_c$$

$$g_c = \frac{\text{ft lb}_m}{\text{lb}_f s^2}$$

$$1 \text{ hp} = 550 \left(\frac{1 \text{ ft}}{1} \times \frac{0.3048 \text{ m}}{\text{ft}} \right) \left(\frac{\dfrac{1 \text{ lb}}{1} \times \dfrac{0.4536 \text{ kg}}{\text{lb}}}{s} \times \frac{32.17 \text{ ft}}{s^2} \times \frac{0.3048 \text{ m}}{\text{ft}} \right)$$

$$1 \text{ hp} = 745.62 \text{ J s}^{-1} = 745.6 \text{ W}$$

The slight difference in value—745.6 and not 745.7—is because of using limited decimals in the conversion values. In most cases, there are very good tables to work from and the conversions as shown in the example were done by someone and collected in tables.

Energy: this is the capacity to do work (J). Energy may be either potential or kinetic. Energy can never be destroyed, but it may be changed from one form to another.

Potential energy: this is latent or stored energy that is possessed by a body due to its condition or position. For example, a 100 kg body of water in a position high above ground, such as in a tank with a free surface 10 m high, will have

$$PE = mhg$$

where m is meter mass flow rate, h is distance fallen in meters, and g is acceleration due to gravity:

$$PE = 100 \frac{\text{kg}}{s} \times 10 \text{ m} \times 9.81 \frac{\text{m}}{s^2}$$

$$PE = 9810 \text{ Js}^{-1} = 9.81 \text{ kW}$$

Potential energy is also released when a product such as coal, oil, or gas is burned. In this case, the solar energy was stored by plants and changed into fossil fuel. Energy can appear in many forms, such as heat, electrical energy, chemical energy, and light. When energy is converted into different forms, there is some loss of utilizable energy:

Energy in = total energy out + stored energy + energy lost

Process design should minimize such "nonproductive" energy losses.

The diesel engine delivers only about 35 to 50 percent of useful energy from the fuel. Modern high-pressure steam generators are only about 40 percent efficient and low-pressure steam boilers are usually only 70 to 80 percent efficient for heating purposes.

Kinetic energy: of a body with mass m moving at velocity V has kinetic energy of

$$KE = \frac{mV^2}{2}$$

If a body of mass m is lifted from ground level to a height of h, then the work done is the force mg multiplied by the distance moved, h, which is mgh. The body has potential energy = mgh.

If the mass is released and starts to fall, the potential energy changes to kinetic energy. When it has fallen the distance h and has a velocity V, then

$$mgh = \frac{mV^2}{2}$$

Kinetic energy for a rotating body is more technical. However, if the mass can be assumed to be concentrated at a point in the radius of rotation, the moment of inertia can be can be used.

Example 2.7

An object with a mass of 1000 kg and moving with a velocity of 15 ms^{-2} at the average radius would be found to have the following kinetic energy:

$$KE = \frac{mV^2}{2\,g}$$

$$KE = \frac{1000 \times 15^2}{2 \times 9.81}$$

$$KE = 11476.89 \text{ J}$$

Kinetic energy is of importance in the movement of all objects in the food processing industry.

Centrifugal force: an object moving around in a circle at a constant tip speed is constantly changing its velocity because of changing direction. The object is subjected to acceleration of $r\omega^2$ or v_t^2/r. Newton's law says that the body will continue in a straight line unless acted upon by a resultant external force called centrifugal force:

$$CF = mr\omega^2 = \frac{mv_t^2}{r}$$

where m = mass, r = radius in meters, and ω = angular velocity = 2π N radians per second.

$$2\pi \text{ rad} = 1 \text{ revolution} = 360°$$

$$1 \text{ rad} = 57.3°$$

A satellite orbiting Earth is continuously falling toward Earth. It just happens that the curvature of Earth allows the satellite to fall exactly as much toward Earth as the Earth's curvature falls away from the satellite. The two bodies therefore stay an equal distance apart.

Centrifugal force is used in separators for the removal of particular matter or separation of immiscible liquids of different densities. The efficiency of a separator can be defined as the number of gravitational forces, where the number of g forces is equal to centrifugal force divided by the gravitational force:

$$CF = \frac{mr\omega^2}{mg} = \frac{r\omega^2}{g}$$

Torque: this is a twisting effect or moment exerted by a force acting at a distance on a body. It is equal to the force multiplied by the perpendicular distance between the line of action of the force and the center of rotation at which it is exerted. Torque (T) is a scalar quantity and measured by the product of the turning force (F) times the radius (r):

$$T = F \times r$$

$$T = Nm$$

2.7 Conversions and Dimensional Analysis

A viscosity table lists viscosity in units of $lb_m/(ft\ s)$. Determine the appropriate SI unit and calculate the conversion factor. The original units have mass (lb_m), distance (ft), and time (s). In SI, the corresponding units should be kilograms, meters, and seconds.

$$kg/ms = (lb_m/fts) \times conversion$$

$$kg/ms = \frac{lb_m}{fts} \times \frac{1 \ kg}{2.2046 \ lb_m} \times \frac{3.281 \ ft}{1 \ m} \times \frac{1 \ s}{1 \ s}$$

$$kg/ms = lb_m/fts \times 1.48866 \ kg/ms$$

But viscosity is also given in Pa s:

$$Pa \ s = (N/m^2) \ s = kg \times m/s^2 \times s/m^2 = kg/ms$$

Problems

1. Set up dimensional equations and determine the appropriate conversion factor to convert one unit of the nonmetric dimensions into the metric dimensions for each of the following:

   a. $ft^2/h = (mm^2/s) \times$ conversion factor

   To solve the problem, look up the conversion for inches to millimeters:[*]

   $$1 \ inch = 25.4 \ mm$$

   $$1 \ ft^2/hr = \frac{1 \ ft \times 1 \ ft}{h} \times \frac{12 \ inch}{1 \ ft} \times \frac{12 \ inch}{1 \ ft} \times \frac{12 \times 25.4 \ mm}{12 \times 1 \ inch} \times$$

   $$\frac{12 \times 25.4 \ mm}{12 \times 1 \ inch} \times \frac{1 \ h}{60 \ min} \times \frac{1 \ min}{60 \ s}$$

   After canceling out, you should get the conversion factor and square millimeters per second.

   b. hundredweight/acre = kilograms per hectare × conversion factor

   Remember that hundredweight is 100 lb. So this can be set up as:

   $$\frac{hundred \ weight}{acre} = \frac{100 \ lb}{acre} \times \frac{0.454 \ kg}{lb} \times \frac{1 \ acre}{4046.86 \ m^2} \times \frac{10000 \ m^2}{ha}$$

   Do the calculation and you will get the answer.

   c. BTUs per cubic foot, degrees Fahrenheit = kilojoules per cubic meter, kelvins × conversion factor

---

[*] I use this format to help guide the students to the answers.

For this conversion, remember that 1 K = 1°C = 9/5°F. This is not a specific temperature—just an interval.

$$\frac{BTU}{ft^3 \, {}^\circ F} = \frac{BTU}{ft^3 \, {}^\circ F} \times \frac{1.055 \text{ kJ}}{BTU} \times \frac{35.3147 \text{ ft}^3}{m^3} \times \frac{{}^\circ F \times \frac{5}{9}}{{}^\circ C}$$

2. Make the following conversions, using standard values for the relationship of length and mass units. You must show the calculations. Using the conversion tools found on the Internet will not help you to understand the process.
   a. 23 miles/h to meters per second

$$\frac{23 \text{ miles}}{h} = \frac{23 \text{ miles}}{h} \times \frac{1609.344 \text{ m}}{\text{mile}} \times \frac{1 \text{ h}}{3600 \text{ m}} = 10.28 \text{ m/s}$$

   b. 35 gal/min to cubic meters per second
      Remember that there are UK gallons and US gallons and that they differ in size.
   c. 35 lb/in.^{2} to kilograms per square meter
   d. 65 lb/in.^{2} to bar
   e. 8500 ft-lb$_f$ to joules and to kilowatt-hours
   f. 10 kW to ft lb$_f$/s and to horsepower

Chapter 3

General Calculations

Many people dread what they will find when they hear words like mathematics, algebra, and calculations. This is very true for most college students and an unfortunate state of affairs when one thinks about it.

Most mathematical problems have an associated reading part and many students tell me that translating words into numerical equations is the most difficult aspect of mathematics. I wish there was some way to make this process easy, but then I also wish that I could change the odds in winning a lottery with equal success.

Calculations are just manipulation of equations and then plugging numbers in to work out on a calculator. I hope that this chapter with the associated appendix will open a new, wonderful world to you. Calculations are easy if you know all the associated rules. I included most of them in Appendix 2. If you are unsure of your grasp of the things that make mathematics easy, look at the stuff in the appendix and place a big sticky note right there so that you can refer to it as often as necessary.

3.1 Basic Principles

Many processes involve adding $(x + y = z)$ and taking away substances $(a - b = c)$ or fractions $(2x - \frac{1}{2} x = 1\frac{1}{2} x)$. In much the same way, energy can be added or removed in the different processes that constitute the system. In many of the processes, the change in physical quantities can be accompanied by chemical reactions and separation processes.

3.2 Conservation of Mass

The basis of mass balances is the conservation of mass. The total material input is equal to the total material output plus any accumulation of material in the equipment. This relationship will apply to the total process, individual components, and any chemical reactions that took place. In the case of a continuous process, the calculations are simplified once the process has reached a steady-state condition. In this case, the accumulation can be regarded as a static quantity and therefore ignored so that input is equivalent to output.

In smaller batch scale operations, the total input is normally removed at some stage so that input is equal to output plus waste (just another form of accumulation). The waste is normally a rather small quantity if we compare it to the relatively large amount of accumulation that can be found in some processes.

3.2.1 Mass Balances

Look at the simultaneous equations in Appendix 2. Mass balances are just simultaneous equations that can be solved by thinking about the problem and using logic. Logic cannot be taught, but it can be developed and, once developed, will be there to use in all future decisions. Mass balances are basically easy and they are good for helping us gain confidence in calculation skills.

To make any headway in a mass balance, the system or part of the system must be defined. It might be necessary to look at different subsystems prior to formulating the total mass balance. An easy method of breaking the system into identifiable sections is to look at the big picture regarding specific components. Putting a boundary around the system will allow us to identify what goes into and what comes out of the bounded system.

For example, when sieving wheat, the amount of wheat that enters the system (feed) must be equal to the wheat that exits the system (product) plus the oversize and undersize material (undesired material) that was separated, plus the dust (undesired material) that was removed, plus any bits and pieces that got stuck in the screens (accumulation):

Dirty wheat = clean wheat + undesired material + accumulation

Any material stream that enters or exits the system must be part of the calculation. This must include any air or vapor transfer. Streams can divide, circulate, or feed into different parts of the system. Keeping track of the mass fractions in the streams is essential to the solution of a problem.

Once the system boundary has been chosen, a basis for the calculation must be chosen. It is possible to use the flow of a specific line per minute, hour, or day. If the flow rate does not change, it is easier to use the flow per time as a volume or mass without thinking of the time. It is, however, much simpler to just use a quantity of material such as 100 kg and then convert this into flow per time at the end of the calculation. It is normally easiest if the compound chosen as the base goes through the system unchanged in mass. When we look at a dehydration process, the water content in the product and the vapor in the air continuously change. All that stays the same is the solid content of the material. It is therefore best to use the solids as the base.

Before starting, try to visualize and interpret the problem. Some people have a knack for these problems and they will find it easy while other people will build the problem-solving skill a bit more slowly. Ponder the problem and then start following the steps. After a while, the visualization process will be quick and natural.

If the following steps are used, calculations become much easier:

- Read the problem at least twice before doing anything else. Follow all the steps listed and read the problem once again while checking the diagram and mass fractions to ensure that everything makes sense.
- Draw a simple diagram of the process and indicate everything that goes in and comes out.
- If a complex system is used, show the boundaries of each subsystem and calculate each subsystem individually.
- Designate symbols for each quantity that needs to be calculated.
- Give the mass fraction (percentage) of each quantity that enters or exits the system.
- Use simultaneous equations to solve the problem (see Appendix 2).

Be sure to use a method that will work for you. Try to visualize the operation as clearly as possible, breaking it down into a logical sequence of events.

Example 3.1

A simple example of a logical sequence of events that has nothing to do with mass balances is found in the manual, "Do an Appendectomy for Idiots":

1. Is the patient on the table? If yes, go to 2; if no, place the patient on the table and go to 1. (Please note the feedback to ensure that the condition is true. Do not jump to the second item after you have given a command. Make sure the command has been properly executed!)
2. Can you see the patient's navel? If yes, go to 3; if no, turn the patient over, and go to 2.
3. Can you see the lower left abdomen? If yes, go to 4; if no, go to 2.
4. Cut the lower left abdomen. Can you see guts? If yes, go to 5; if no, go to 4.
5. Find the appendix, rip it out, cut it off, put it in a bottle of alcohol to make a paperweight, sew up the abdomen after counting swabs and hardware, etc.

A simple mass balance is certainly easier than an appendectomy, so get started.

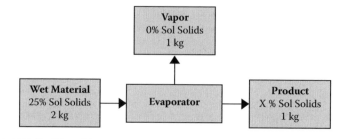

Figure 3.1 Mass balance setup for evaporation problem.

Example 3.2

An aqueous solution contains 25 percent (m/m) soluble solids. After evaporation, the solution lost half of the original mass (Figure 3.1). What is the concentration of soluble solids in the concentrated solution? You can use any mass as long as the final mass is half of the original mass and the amount of water vapor removed is the same as the final mass. In this case, it is easy to use 2 and 1 but 10 and 5 would also work fine.

Mass balance is

$$2 \text{ kg(wet)} = 1 \text{ kg(vapor)} + 1 \text{ kg(concetrate)}$$

Soluble solids balance is

$$2 \text{ kg} \times 0.25 = 1 \text{ kg} \times 0 + 1 \text{ kg} \times X$$

$$0.5 = X$$

The concentrate will contain 50 percent soluble solids. Think about it. The original 2 kg contained 500 g soluble solids and 1500 g water for a final mass of 2000 g. To lose half the mass, there are 500 g soluble solids in 500 g water, or 50 percent soluble solids. This is a simple problem, but one in which you can check your answer with simple logic. Checking the relative amounts of all the compounds is the most important part of mass balances.

You could also set the problem up as a loss function:

$$1 \text{ kg} \times X = 2 \text{ kg} \times 0.25 - 1 \text{ kg} \times 0$$

Soluble solids balance is

$$1 \text{ kg} \times X = 2 \text{ kg} \times 0.25 - 1 \text{ kg} \times 0$$

$$X = 0.5$$

Example 3.3

A company produces a new Asian style condiment that will contain 2 percent (m/v) salt and 6 percent (m/v) sugar in the final product. The salt and sugar are

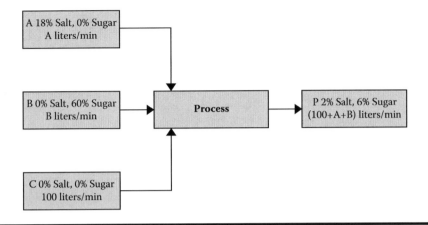

Figure 3.2 Mass balance setup for the condiment problem.

added as solutions that are fed into the main pipeline. If the salt solution contains 18 percent (m/v) salt and the sugar solution contains 60 percent (m/v) sugar (see Figure 3.2), how much of each of the concentrates must be added to the condiment stream if it flows at 100 L per minute? Since the flow is given in liters per minute, we can use just the liters.

Overall mass balance is

$$\text{Condiment mix } (C) + \text{salt } (A) + \text{sugar } (B) = \text{product } (P)$$

Salt balance is

$$100 \times 0.0 + A \times 0.18 + B \times 0.0 = (100 + A + B) \times 0.02$$

Sugar balance is

$$100 \times 0.0 + A \times 0.0 + B \times 0.6 = (100 + A + B) \times 0.06$$

Solving for two unknowns, we have the required two equations:

$$0.18A = 2 + 0.02A + 0.02B$$

$$0.16A - 0.02B = 2 \rightarrow (1)$$

$$0.60B = 6 + 0.06A + 0.06B$$

$$-0.06A + 0.54B = 6 \rightarrow (2)$$

To solve this, one term in both equations must have the same value. Multiplying Equation (1) with 27 will ensure that the values for B are the same in both equations. Adding the two equations will eliminate B and we can solve for A; then the answer is used to solve for B.

$$27 \times 0.16A - 27 \times 0.02B = 27 \times 2$$

$$4.32A - 0.54B = 54 \rightarrow (3)$$

Adding Equation 2 to Equation 3 gives

$$4.26A = 60$$

$$A = \frac{60}{4.26} = 14.1$$

Substituting in Equation 2 gives

$$-0.06 \times 14.1 + 0.54B = 6$$

$$0.54B = 6 + 0.846$$

$$B = \frac{6.846}{0.54} = 12.68$$

The process requires that 14.1 L/min of the salt solution and 12.68 L/min of the sugar solution be added. The new flow rate will be 126.8 L/min. It should contain 6 percent sugar; thus, 126.8×0.06 is 7.61 kg of sugar and 2.54 kg of salt.

How much sugar is in 12.7 L of a 60 percent solution? There are 60 kg per 100 L; thus, 0.6×12.68 kg is 12.68 L. This is equal to 7.61 kg.

Calculating the salt in the same way, we get 2.54 kg of salt in 14.1 L of solution. The answer is verified. (Remember that answers can vary slightly because of rounding up or down.)

Example 3.4

This example will expand the scope of our mass balance calculations. The same principles are presented in a deluxe version.

A winery is making a new fruit-flavored blended mix. The final product contains 25 percent clear peach juice. The wine portion can be blended from A, B, or C. The alcohol and sugar levels for the wine and the blend are given in Table 3.1.

Use 100 kg as a basis for the blending operation. With that, one can easily convert to larger quantities.

Total mass balance is

$$A + B + C + J = 100$$

Alcohol balance is

$$0.143A + 0.169B + 0.153C + 0.0J = 100 \times 0.12$$

Sugar balance is

$$0.037A + 0.024B + 0.054C + 0.13J = 100 \times 0.055$$

Table 3.1 Alcohol and Sugar Levels for the Wines and the Blend

Ingredient	Alcohol % (m/m)	Sugar % (m/m)
Wine A (A)	14.3	3.7
Wine B (B)	16.9	2.4
Wine C (C)	15.3	5.4
Peach Juice (J)	0	13
Blend (P)	12	5.5

There are four unknowns in the problem, but only three equations; the problem cannot be solved. Read the problem again: We need to use 25 percent peach juice; therefore, J is equal to 25 kg and the three equations can be rewritten using 25 in place of J.

Total mass balance is

$$A + B + C = 100 - 25 = 75$$

Alcohol balance is

$$0.143A + 0.169B + 0.153C = 12$$

Sugar balance is

$$0.037A + 0.024B + 0.054C = 5.5 - 25 \times 0.13 = 2.25$$

Multiply the total mass balance with 0.143 and subtract the answer from the alcohol balance to remove A. Multiply the total mass balance with 0.037 and subtract the answer from the sugar balance, again eliminating A. This gives two equations with two unknowns:

$$0.026B + 0.01C = 1.275$$

$$-0.013B + 0.017C = -0.525$$

Multiplying the second equation with 2 and then adding it to the first equation will eliminate B:

$$-0.026B + 0.034C = -1.05$$

$$0.044C = 0.225$$

$$C = 5.11 \text{ kg}$$

Substitution of B in one of the formulas will solve for B:

$$0.026B + 0.01 \times 5.11 = 1.275$$

$$B = 47.07 \text{ kg}$$

Substituting B and C in the total mass balance gives A:

$$A = 22.82 \text{ kg}$$

The blend will be made up of 22.82 kg A, 47.07 kg B, 5.11 kg C, and 25 kg peach juice. To check the answers, use the alcohol balance:

$$0.143 \times 22.82 + 0.169 \times 47.07 + 0.153 \times 5.11 = 12$$

$$3.263 + 7.955 + 0.782 = 12$$

You might verify the sugar the sugar balance in the same way.

For those of you that want to use Solver in Excel, the problem can be set up as in Table 3.2.

Using the amount of product to produce and forcing the target to a set value of 100 kg, all that is required is to make sure that the quantity of ingredients produced is not negative. In mathematics, negative numbers are very useful. In real manufacturing situations, it means that you have to take stuff out of the product—not an easy task. The other constraints are the amount of alcohol and sugar in the final product. This very simple problem can be done in about 4 minutes on Solver (see Table 3.3). If you have Microsoft Office and Excel, and if solver is not available in the Tools menu, then just go to Tools and find Add-Ins; Solver will be there.

Example 3.5

One ton of sliced carrots is dehydrated in a concurrent flow dehydrator from 85 percent (m/m) to 20 percent (m/m) moisture content. Air with moisture content of 0.012 kg water per kilogram of dry air enters the dehydrator at a rate of 500 kg dry air per kilogram of dry solid (see Figure 3.3). What is the moisture content of the air leaving the system?If the solids entering contain 85 percent (m/m) water, that means 85 g water and 15 g solids in 100 g material. The amount of water entering the dehydrator with 1000 kg is

$$1000 \times 0.85 = 850 \text{ kg}$$

The amount of solids is thus 150 kg. But the final solids also contain 20 percent (m/m) moisture; therefore, the final mass is

$$\frac{150}{0.8} = 187.5 \text{ kg}$$

The amount of water removed per hour from each ton is

$$850 \text{ kg} - 37.5 \text{ kg} = 812.5 \text{ kg}$$

Table 3.2 Problem Setup in Excel Solver

	A	B	C	D	E	F
1		A	B	C	D	P
2	Alcohol	14.3	16.9	15.3	0	12
3	Sugar	3.7	2.4	5.4	13	5.5
4	Amount	22.814685	47.071679	5.113636	25	= SUM(B4:E4)
5	Alcohol	= B4*B2/100	= C4*C2/100	= D4*D2/100	= E4*E2/100	= SUM(B5:E5)
6	Sugar	= B4*B3/100	= C4*C3/100	= D4*D3/100	= E4*E3/100	= SUM(B6:E6)

Table 3.3 Answer Report for the Excel Solver Solution

Cell	Name	Original Value	Final Value
Target Cell (Value of)			
F4	Amount P	100.000	100.000
Adjustable Cells			
B4	Amount A	22.815	22.815
C4	Amount B	47.072	47.072
D4	Amount C	5.114	5.114
Constraints			
F6	Sugar P	5.500	F6 = F3
F5	Alcohol P	12.000	F5 = F2
B4	Amount A	22.815	B4 ≥ 0
C4	Amount B	47.072	C4 ≥ 0
D4	Amount C	5.114	D4 ≥ 0

Note: The cell references refer to the previously given Excel sheet.

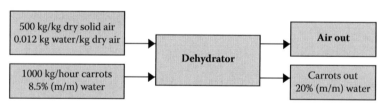

Figure 3.3 Mass balance setup for the dehydrator problem.

The amount of air entering the system is

$$150 \text{ kg} \times \frac{500 \text{ kg}}{\text{kg}} = 75000 \text{ kg}$$

The moisture in the incoming air is

$$75000 \text{ kg} \times \frac{0.012 \text{ kg}}{\text{kg}} = 900 \text{ kg}$$

The total moisture leaving the system is

$$812.5 \text{ kg} + 900 \text{ kg} = 1712.5 \text{ kg}$$

Moisture content of the air leaving the system is

$$\frac{1725.5 \text{ kg}}{75000 \text{ kg}} = 0.023 \text{ kg moisture/kg dry air}$$

There are other ways of dealing with the same problem.

The moisture entering the system with a kilogram of dry material is given by

$$\frac{0.85}{0.15} \times 1 \text{ kg} = 5.67 \text{ kg}$$

The moisture leaving the system with the "dry" product is

$$\frac{0.20}{0.80} \times 1 \text{ kg} = 0.25 \text{ kg}$$

The moisture entering the system with the air is

$$500 \times 0.012 \text{ kg} = 6.0 \text{ kg}$$

Since water in must equal water out, the water uptake by the air is

$$5.67 \text{ kg} + 6 \text{ kg} - 0.25 \text{ kg} = 11.42 \text{ kg}$$

The moisture content of the air leaving is thus

$$\frac{11.42 \text{ kg}}{500 \text{ kg}} = 0.023 \text{ kg moisture/kg dry air}$$

There are many different ways to tackle most problems; find out what works for you and keep the logic working. Be very cautious regarding the percentage composition of substances. The percentage can be given on a dry mass basis or a wet mass basis, giving very different results.

If 20 g of a pure substance is mixed with 40 g water, the total mass is 60 g. On a wet mass basis, the moisture content will be

$$\frac{40}{60} \times 100 = 66.67\%$$

And on a dry mass basis, the moisture content of the wet mixture will be

$$\frac{40}{20} \times 100 = 200\%$$

This example is certainly silly but it is given to show the effect of the base used for calculations.

Example 3.6

We use crystallization as a way to purify some soluble chemical compounds. If we dissolve 100 kg of refined sugar that contains 94 percent (m/m) sucrose, 3 percent (m/m) water, and 3 percent soluble solids that will not crystallize in 25 kg water at 80°C, we should have an almost saturated solution.

Calculate the percent sucrose in the solution:

Total sucrose = 94 kg

Total water = 3 + 25 kg (3 kg water in the sugar and 25 kg that we dissolve it in)

Total mass = 100 + 25 kg

$$\frac{100 \times 0.94}{100 + 25} \times 100 = 75.2\% \text{ (m/m)}$$

This solution will be saturated at 78 percent sucrose. If we cool the solution to 20°C where a saturated solution contains 67 percent (m/m) sucrose, how much sucrose will crystallize?

If we assume that the inert soluble solid will not compete with the sucrose for water molecules, we can calculate how much of the sucrose will stay in solution:

$$125 \times 0.67 = 83.75 \text{ kg}$$

The difference between this amount and the total amount that went in is the amount that will crystallize:

$$94 - 83.75 = 10.25 \text{ kg}$$

After centrifugation, the crystals contain 10 percent of the supernatent. What is the mass of crystals and mother liquor that will exit the centrifuge?

$$\frac{10.25}{0.9} = 11.39 \text{ kg}$$

What will the final mass of sugar be after the crystals are dried to 0 percent moisture?

First one should calculate the amount of moisture, sugar, and soluble solids in the adhering moisture.

Sugar content is

$$1.24 \times 0.67 = 0.83 \text{ kg}$$

Water and soluble solids are

$$1.24 - 0.83 = 0.41 \text{ kg}$$

In the original 125 kg we had 3 kg soluble solids that will not crystallize. After removal of 10.25 kg pure sucrose, we had 3 kg soluble solids in 114.75 kg total soluble phase. The concentration of the soluble solid is thus

$$\frac{3}{114.75} \times 100 = 2.61\%$$

The amount of the soluble solid in the crystals is

$$1.24 \times 0.0261 = 0.032 \text{ kg}$$

This amount of soluble solid will adhere to the recrystallized sucrose. What will the percentage contamination be after drying?

Solids amount is

$$10.25 + 0.032 = 10.282 \text{ kg}$$

The amount of soluble solid in the sucrose is now

$$\frac{0.032}{10.282} \times 100 = 0.31\% \text{ (m/m)}$$

The sucrose is thus more than 99.69 percent pure. Recrystallization is the way in which substances that can crystallize are purified.

Problems

1. What is the final salt concentration if 20 kg of a solution containing 21 percent (m/m) salt is mixed with 10 kg of a solution containing 12 percent (m/m) salt?

Salt mass: = 20 kg × 0.21 kg/kg + 10 kg × 0.12 kg/kg = 4.2 kg + 1.2 kg = 5.4 kg

Total solution mass: = 20 kg + 10 kg = 30 kg

Now calculate the final concentration.

2. How much water is removed when 100 kg of a solution containing 11 percent (m/m) total solids has to be concentrated to produce a liquid containing 35 percent (m/m) total solids?

Total mass balance is

$$\text{Original } (O) - \text{water } (W) = \text{final } (F)$$

Moisture balance is

$$100 \text{ kg} \times (1.00 - 0.11) - W = (100 - W)\text{kg} \times (1.00 - 0.35)$$

3. While making ketchup, the tomato concentrate is flowing through a pipeline at 50 kg/min. At what rate must a 23 percent (m/m) solution of salt be added to ensure 3 percent salt in the final product?
   Total mass balance is

$$\text{Concentrate} + \text{Salt} = \text{Product}$$

$$50 \text{ kg/min} + S \text{ kg/min} = (50 + S)\text{kg/min}$$

Salt balance is

$$0 \times 50 \text{ kg/min} + 0.23 \, S \text{ kg/min} = 0.03 \times (50 + S)\text{kg/min}$$

4. A brewer's spent grain contains 80 percent (m/m) moisture. When the grain has lost half of its weight, what is the moisture content?
5. Calculate the quantity of dry air that is required to dry 50 tons of food from 19 to 11 percent moisture. The air enters with a moisture content of 0.003 kg water/kg of dry air and exits with a moisture content of 0.009 kg water/kg of dry air.
6. Orange juice concentrate 3:1 is reconstituted by mixing one volume of concentrate with 3 volumes of water. Consumers have indicated that this tastes artificial and manufacturing is producing a new 4:1 concentrate. Before packaging, this will be blended with fresh juice to produce a 3:1 concentrate. If you want to produce 1000 kg of the new product, how much 4:1 concentrate must be prepared and how much juice must be kept in reserve for dilution purposes?
7. To produce sausage formulations, packers normally use lean beef, pork belly, and soy concentrate. The manufacturer uses 3 percent soy in final sausage mass. Water is added to the formulation as ice. The final sausage must contain: 15 percent protein, 60 percent moisture, and 25 percent fat.
   The ingredients contain the following:

Ingredient	Protein	Moisture	Fat
Lean beef	20%	67%	13%
Pork belly	10%	40%	50%
Soy	90%	7%	0%

How much of each ingredient should be combined to make 600 kg of sausage emulsion?

Chapter 4

Properties of Fluids

A fluid is a substance that cannot maintain its own shape and has no rigidity. It will flow and alter its shape to fill the container that retains it. It differs from a solid in that it suffers deformation due to shear stress, however small the shear stress may be. The only criterion is that sufficient time should elapse for the deformation to take place. In this sense, a fluid is shapeless.

Fluids may be divided into liquids and gases. A liquid is only slightly compressible and there is a free surface when it is placed in an open vessel. On the other hand, a gas always expands to fill its container. A vapor is a gas that is near the liquid state. It is easier to thinks of gases as compressible and fluids as incompressible.

Water will be the liquid that will be used as an example. It is a well-known commodity and one that can be used as an example for many problems. Water may contain up to 3 percent of air in solution, which at subatmospheric pressures tends to be released. Provision must be made for this when designing pumps, valves, pipelines, etc.

4.1 Physical Properties of Flow

The principal physical properties of fluids are described as follows.

4.1.1 Pressure

Pressure is force acting on a specific area. If the force is measured in newtons and the area in square meters, then the pressure is in pascals (Pa):

$$Pa = \frac{N}{m^2}$$

The pressure caused by a column of liquid acts equally in all directions and depends upon the depth and density of the liquid. It is thus defined as force per unit area and is expressed as weight-force and area or the height of a column of liquid that will produce a similar pressure at its base:

$$\text{Pressure (Pa)} = \text{Force}\left(\frac{N}{kg}\right) \times \text{head (h)} \times \text{density (}\rho\text{)}$$

$$\text{Pressure (Pa)} = \frac{N}{kg} \times m \times \frac{kg}{m^3}$$

Pressure can also be expressed as the column of liquid (h) times the density of the liquid (ρ) multiplied by the gravitational constant (g):

$$\text{Pressure (Pa)} = \rho g h$$

$$\text{Pressure (Pa)} = \text{density}\left(\frac{kg}{m^3}\right) \times \text{gravity}\left(\frac{m}{s^2}\right) \times \text{head (m)}$$

Example 4.1

A tank is filled with water to a depth of 2.5 m. What is the pressure at the bottom of the tank due to the water alone?

$$P = \rho g h$$

$$P = 10^3 \frac{kg}{m^3} \times 9.81 \frac{m}{s^2} \times 2.5 \text{ m}$$

$$P = 2.45 \times 10^4 \text{ Nm}^{-2} = 2.45 \times 10^4 \text{ Pa} = 24.5 \text{ kPa}$$

At any point below the surface of a liquid the pressure is equal to pressure on the surface P_s, plus the pressure of the column of liquid above it, to give

$$P_A = P_S + \rho g h$$

Example 4.2

What is the total pressure at the bottom of the tank if the surface of the water in the tank is at atmospheric pressure? The normal value for atmospheric pressure is 101.3 kPa.

$$P_A = P_S + \rho g h$$

$$P_A = 101.3 \text{ kPa} + 24.5 \text{ kPa} = 125.8 \text{ kPa}$$

4.1.2 Density (ρ)

The density of a fluid is its mass per unit volume. In the SI system, it is expressed as kilograms per cubic meter. Water is at its maximum density of 1000 kg/m³ at 4°C. There is a slight decrease in density with increasing temperature, but for practical purposes the density of water is 1000 kg/m³. Relative density is the ratio of the density of a liquid to that of water.

4.1.3 Specific Mass (w)

The specific mass of a fluid is its mass per unit volume. In the SI system, it is expressed in newtons per cubic meter. At normal temperatures, water is 9810 N/m³ or 9.81 kN/m³ (approximately 10 kN/m³ for ease of calculation).

4.1.4 Specific Gravity (SG)

The specific gravity of a fluid is the ratio of the mass of a given volume of liquid to the mass of the same volume of water. Thus, it is also the ratio of a fluid density to the density of pure water, normally all at 15°C.

4.1.5 Bulk Modulus (k)

For practical purposes, liquids may be regarded as incompressible. However, there are certain cases, such as unsteady flow in pipes, where the compressibility should be taken into account. The bulk modulus of elasticity, k, is given by

$$k = P / \ V/V = \frac{P}{\dfrac{V}{V}} = \frac{PV}{V}$$

where P is the increase in pressure that, when applied to a volume V, results in a decrease in volume V. Since a decrease in volume must be associated with a proportionate increase in density, the equation may be expressed as

$$k = P / \ \rho/\rho = \frac{P\rho}{\rho}$$

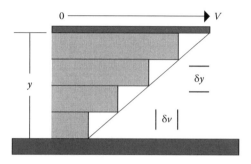

Figure 4.1 **Viscous deformation.**

For water, k is approximately 2150 Nm/m² at normal temperatures and pressures. It follows that water is about 100 times more compressible than steel.

4.1.6 Ideal Fluid

An ideal or perfect fluid is one in which there are no tangential or shear stresses between the fluid particles. The forces always act normally at a section and are limited to pressure and acceleration forces. No real fluid fully complies with this concept, and for all fluids in motion there are tangential stresses present that have a dampening effect on the motion. However, some liquids, including water, are near to an ideal fluid, and this simplified assumption enables mathematical or graphical methods to be adopted in the solution of certain flow problems.

4.1.7 Viscosity

Fluids are substances that undergo continuous deformation when shear stress is applied. Viscosity* of a fluid is a measure of its resistance to tangential or shear stress. It arises from the interaction and cohesion of fluid molecules. All real fluids possess viscosity, though to varying degrees. The shear stress in a solid is proportional to strain, whereas the shear stress in a fluid is proportional to the rate of shearing strain. It follows that there can be no shear stress in a fluid that is at rest.

Consider a fluid confined between two plates that are situated a very short distance δy apart. The lower plate is stationary while the upper plate is moving at velocity v. The fluid motion is assumed to take place in a series of infinitely thin layers or laminae, free to slide one over the other. There is no cross flow or turbulence (see Figure 4.1).

The layer adjacent to the stationary plate is at rest while the layer adjacent to the moving plate has a velocity v. The rate of shearing strain or velocity gradient $\gamma = dv/dy$. The dynamic viscosity or, more simply, the viscosity μ is given by

---

* Viscosity describes a liquid's resistance to flow. High viscosity means a liquid with flow properties that are near to those of a solid (i.e., glass). Viscosity is the reciprocal of fluidity.

$$\mu = \frac{\tau}{\gamma}$$

where μ = shearing stress/rate (τ) of shearing strain (dv/dy) so that

$$\mu = \frac{\tau}{\dfrac{dv}{dy}}$$

so that

$$\tau = \mu \frac{dv}{dy}$$

The SI units for viscosity are kg/(m•s) or Pa•s (pascal second). The cgs unit for viscosity is poise. One Pa equals 10 poise or 1000 centipoise (cP)

Dynamic viscosity – pascal second = Pa × s = μ

This expression for the viscous stress was first postulated by Newton and is known as Newton's equation of viscosity. Fluids that exhibit a linear increase in shear stress with the rate of shearing strain are called Newtonian fluids. The slope or coefficient of proportionality, μ, in the equation is constant for Newtonian fluids. Gases and low molecular weight liquids are normally Newtonian.

In many problems concerning fluid motion, the viscosity appears with the density in the form

$$\nu \equiv \frac{\mu}{\rho}$$

(independent of force) and it is convenient to employ a single term ν known as the kinematic viscosity.

Kinematic viscosity = Area per second = $\dfrac{m^2}{s} = \nu$

All fluids that do not exhibit a linear relationship between shear stress and rate of shearing are non-Newtonian fluids. For non-Newtonian fluids, the viscosity, defined by the ratio of shear stress to shear rate is called apparent viscosity. Non-Newtonian fluids exhibit shear-thickening or shear-thinning behavior and some exhibit yield stress (see Figure 4.2).

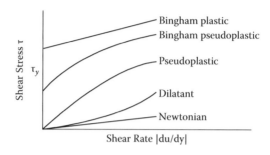

Figure 4.2   Shear diagrams.

The apparent viscosity or consistency of a non-Newtonian fluid will change with a change in the rate of shear. The liquid materials that we process in the food industry, such as tomato sauce, mayonnaise, starch slurries, and suspensions, are non-Newtonian.

The non-Newtonian fluids can be classified in shear-thinning fluids or pseudo-plastic fluids and shear-thickening or dilatant fluids.

4.1.8 Temperature Effects on Fluid Flow

All liquids have a reduced resistance to flow when the temperature is increased. There is normally a 2 percent change in viscosity or apparent viscosity with every degree Celsius change in temperature. For some substances, the change is much greater; pitch shows a 30 percent decrease in viscosity when the temperature changes from 20°C to 21°C.

It is very important that temperature should be controlled when measuring viscosity. Since the ease of flow and thus pumping will be influenced by temperature, it is just as important to specify the temperature of a specific liquid that the system is designed to pump. The empirical equation used most frequently to relate changes in viscosity with changes in temperature has been experimentally determined as

$$\log \mu = \frac{B}{T} + C$$

where T is absolute temperature, and B and C are constants for the fluid.

Gases normally have an increase in viscosity when the temperature increases.

4.1.9 Shear-Thinning Fluids

4.1.9.1 Bingham Plastics

Bingham plastic fluids exhibit a linear increase in shear stress with the rate of shearing strain just like Newtonian fluids. Flow is initiated by starting pressure of force called the yield value. Once the yield value is reached, the slope or coefficient of proportionality, μ, in the equation is constant, as for Newtonian fluids. The rheogram of this material will have a constant slope once the yield value is reached.

$$\tau = \tau_y + \mu_\infty \gamma$$

Tomato sauce will not start to flow from the bottle if it is carefully inverted. If the bottle is struck, the sauce will flow freely under gravity. Examples of Bingham plastics include chocolate liquor, hot tomato sauce, and cream style sweet corn.

4.1.9.2 Pseudoplastics

Shear-thinning fluids have slopes that decrease with an increase in shear rate. These fluids used to be referred to as pseudoplastic. The apparent viscosity for such material is

$$\mu = K\gamma^{n-1}$$

The coefficient K (kg/(m·s^{2-n})) is the consistency index and n is the power law exponent. The exponent n is dimensionless. The apparent viscosity has the same units as viscosity, although the value varies with the rate of shear. It is very important to report the shear rate with the observed reading.

Shear-tinning fluids are characterized by the phenomenon that there is never a linear relationship between the rate of shear and the shearing force. If the material requires a yield value to initiate flow, it is classified as a Bingham pseudoplastic or Herschel-Bulkley plastic material. Many emulsions show this effect.

4.1.10 Shear-Thickening Materials

Shear-thickening or dilatant materials show an increase in apparent viscosity (thickening) as the rate of shear increases.

This phenomenon can cause considerable problems when dilatant materials are pumped. As the pumping rate increases, the material becomes solid within the pump. Heavy starch suspensions, candies, milk chocolate filled with buttermilk powders, and certain kinds of honey all exhibit dilatant flow behavior. As soon as the agitation stops, most of the dilatant liquids will return to their original consistency.

A possible explanation for this behavior is that in the resting state the interparticular spaces are small. There is enough water present to lubricate the particles and enhance flow. As soon as the liquid is agitated, the spaces increase in size and the amount of fluid is insufficient to lubricate the particles fully.

4.2 Surface Tension and Capillary Action

Cohesion is the attraction that similar molecules have for each other. *Adhesion* is the attraction that dissimilar molecules have for each other. *Surface tension* is the physical property of liquids in which the exposed surface tends to contract to the smallest possible area. It is this property that enables a drop of water to be held in suspension from the lip of a tap, a vessel to be filled with liquid slightly above the brim and yet not spill, or a needle to float on the surface of a liquid. It is as though the surface consists of an elastic membrane, uniformly stressed, that tends always to contract the superficial area. Thus, we find that bubbles of gas in a liquid and droplets of moisture in the atmosphere are approximately spherical in shape.

If one of the fluids is the vapor phase of the liquid being tested, the measurement is referred to as surface tension. If the surface investigated is the interface of two liquids, the measurement is referred to as interfacial tension. In either case, the more dense fluid is referred to as the *heavy phase* and the less dense fluid is referred to as the *light phase.*

The surface tension force across any imaginary line at a free surface is proportional to the length of the line and acts in a direction perpendicular to it. The surface tension per unit length is expressed in millinewtons per meter (mN/m) or dynes per centimeter. These units are equivalent so that 1.00 dynes/cm = 1 mN/m. This excess energy that exists at the interface of two fluids has a small magnitude of about 73 mN/m for water in contact with air at room temperature. There is a slight decrease in surface tension with increasing temperature (see Table 4.1). Water at ambient temperature has a high surface tension in the range of 72 dynes/cm while alcohols are in a low range of 20 to 22 dynes/cm. Solvents, typically, are in the 20 to 30 dynes/cm range. Liquids with high surface tension values also have high latent heat values. If any formulation changes at the molecular level, then the surface tension will change. If the formulation changes due to the addition of another chemical or the addition of a surfactant, or if anything contaminates the fluid in question, the surface tension will change.

In most applications in hydraulics, surface tension is of such small magnitude compared to the other forces in action that it is normally ignored.* Surface tension is only of importance where there is a free surface and the boundary dimensions are small. Thus, in the case of hydraulic models, surface tension effects, which are of no consequence in the prototype, may influence the flow behavior in the model, and this source of error in simulation must be taken into consideration when interpreting the results.

Surface tension effects are very pronounced in the case of tubes of small bore open to the atmosphere. These may take the form of manometer tubes in the laboratory or open pores in the soil. For instance, when a small glass tube is dipped

---

* Newton's second law: if several forces act on a body, it will obey each as though the others did not exist.

Table 4.2 Secondary Energy or Frictional Loss in Pipelines Expressed as Equivalent Length of Pipe in Meters

Diameter (mm)	Ball Valve Fully Open	Gate Valve Fully Open	Foot Valve Fully Open	Nonreturn Valve	Sieve	90° Elbow	45° Elbow	Tee Piece	Entrance into Pipe	Exit from Pipe
15	4.88	0.11	1.22	1.16	3.05	0.55	0.24	1.04	0.46	0.27
20	6.40	0.14	1.52	1.39	3.66	0.69	0.31	1.37	0.61	0.37
25	8.23	0.18	1.83	1.58	4.27	0.84	0.41	1.77	0.76	0.46
32	11.28	0.24	2.13	2.74	4.88	1.14	0.52	1.79	1.07	0.61
40	13.72	0.29	2.44	3.35	5.49	1.36	0.61	2.74	1.22	0.73
50	16.76	0.38	2.74	4.27	6.10	1.62	0.76	3.66	1.52	0.91
65	19.81	0.43	3.05	5.18	6.71	1.98	0.91	4.27	1.83	1.10
80	25.91	0.53	3.66	5.79	7.62	2.50	1.16	4.88	2.29	1.37
90	28.96	0.61		6.71		2.99	1.37	5.49	2.74	1.59
100	36.58	0.73		7.67		3.66	1.52	6.71	3.05	1.83
125	44.20	0.88		10.06		4.27	1.98	8.23	3.96	2.23
150	51.82	1.07		12.19		4.88	2.29	10.06	4.57	2.74

into water, it will be found that the water rises inside the tube. The phenomenon is known as capillary action, and the tangential contact between the water and the glass indicates that the internal cohesion of the water is less than the adhesion between the water and the glass. The water surface in the tube, or meniscus as it is called, is concave. The pressure of the water within the tube adjacent to the free surface is less than atmospheric.

4.2.1 Vapor Pressure

Liquid molecules that possess sufficient kinetic energy are projected out of the main body of a liquid at its free surface and pass into the vapor. The pressure exerted by this vapor is known as the vapor pressure, P_v. An increase in temperature is associated with a greater molecular agitation and therefore an increase in vapor pressure. When the vapor pressure is equal to the pressure of the gas above it, the liquid boils. The vapor pressure of water at 15°C is 1.72 kPa (1.72 kN/m²).

4.2.2 Atmospheric Pressure

The pressure of the atmosphere at the Earth's surface is measured by a barometer. At sea level the atmospheric pressure averages 101 kPa and is standardized at this value. There is a decrease in atmospheric pressure with altitude; for instance, at 1500 m it is reduced to 88 kPa. The water column equivalent has a height of 10.3 m at sea level and is often referred to as the water barometer. The height is hypothetical, since the vapor pressure of water would preclude a complete vacuum being attained. Mercury is a much superior barometric liquid, since it has a negligible vapor pressure. In addition, its high density results in a column of reasonable height—about 0.75 m at sea level.

As most pressures encountered in hydraulics are above atmospheric pressure and are measured by instruments that record relatively, it is convenient to regard atmospheric pressure as the datum (i.e., zero). Pressures are then referred to as gauge pressures when above atmospheric and vacuum pressures* when below it.

Gauge pressure = absolute pressure − atmospheric pressure

If true zero pressure is taken as datum, the resulting pressures are said to be absolute and can never be negative.

---

* Never refer to vacuum as a negative pressure. Pressure is force/area and force is mass × acceleration, both positive, so a negative pressure can only result if a negative area exists. To make a force negative, an opposing force, acting in an opposite direction, will have to be employed. It is only when using a datum that is not at zero that a pressure can be compared to another pressure and, if one pressure is lower, a negative answer may result. The negative means a pressure is lower than the pressure at the datum; it is still not a negative pressure. Some students even refer to negative vacuum, whatever that is.

4.3 Hydrostatics

Hydrostatics is the branch of fluid mechanics that is concerned with fluids at rest. No tangential or shear stress exists between stationary fluid particles. In hydrostatics, all forces act normally to a boundary surface and are independent of viscosity. The laws that control the system are fairly simple, and analysis is based on a straightforward application of the mechanical principles of force and momentum. Solutions are exact with no need to have recourse to experimentation.

4.3.1 Pressure Intensity

The pressure intensity or, more simply, the pressure on a surface is the pressure force per unit area. It is interesting to note that the pressure of a liquid at the surface is proportional to the area of the surface.

Example 4.3

If we use the illustration in Figure 4.3 where the area of M = 10 m², the area of S = 1 m², and the force on S is 1 kg, what mass will be supported on M?

$$\frac{F_a}{A_a} = \frac{F_b}{A_b}$$

$$F_a = \frac{F_b \times A_a}{A_b} = \frac{1 \times 10}{1} = 10 \text{ kg}$$

A mass of 1 kg on S will support a mass of 10 kg on M.

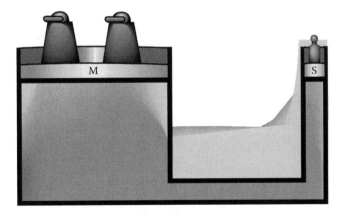

Figure 4.3 **Pressure exerted by a liquid on a surface is proportional to the area of the surface.**

Example 4.4

What is the height X of the water column if the locomotive in Figure 4.4 has a mass of 50,000 kg, the area of the plunger is 500 cm², and the area of the water column is 1 cm²?

The water pressure at the bottom of the column of water must be

$$F_a = \frac{F_b \times A_a}{A_b} = \frac{50,000 \times 1}{500} = 100 \text{ kg} \times 9.81 \text{ m/s}^2 = 981 \text{ N/cm}^2$$

$$P = \rho g h = 9810000 \text{ Pa}$$

$$h = \frac{9810000 \text{ Pa}}{9.81 \text{ m/s}^2 \times 1.0 \times 10^3 \text{ kg/m}^3} = 100 \text{ m}$$

If the water column has a surface area of 1 cm² and a length of 100 m = 100 × 100 cm, then the volume of the water is

$$V = 1 \text{ cm}^2 \times 100 \times 100 \text{ cm} = 10000 \text{ cm}^3$$

It will thus require 10 L of water to balance the locomotive!

The pressure of the liquid at any point below the surface is proportional to the depth of the point below the surface. The vertical downward pressure force acting on the horizontal lamina is equal to the mass of the column of fluid vertically above it, plus the pressure intensity at the interface with another fluid. For static equilibrium, there must be a corresponding upward vertical pressure below the lamina. In the case of an incompressible liquid in contact with the atmosphere, the gauge pressure (in pascals) is given by

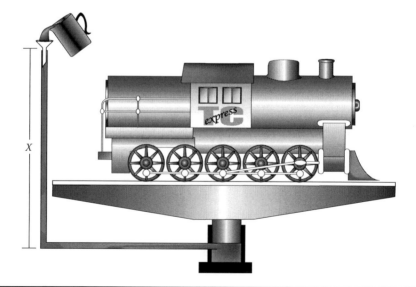

Figure 4.4 A quantity of water can be made to balance any mass.

$$P = wh = \rho gh$$

where w is the specific mass of the liquid and h is the depth below the free surface. The latter is referred to as the pressure head and is generally stated in meters of liquid. The form of the equation shows that the pressure increases linearly with depth. As gravity is the physical property that is involved, the free surface of a still liquid is always horizontal and the pressure intensity is the same on any horizontal plane within the body of the liquid.

4.4 Basic Concept of Fluid Motion

4.4.1 Introduction

As previously stated, exact mathematical solutions for the forces exerted by fluids at rest can be readily obtained since only simple pressure forces are involved in hydrostatics. When a fluid in motion is considered, the problem of analysis becomes much more complex.

As a fluid flows along a straight horizontal pipe, frictional losses occur that result in loss of pressure energy. When one measures the pressure drop for different flow rates, the relationship between flow rate and pressure drop will be linear for part of the relationship and then it will become nonlinear. At the point where the relationship becomes nonlinear, there is a change in the nature of the flow (see Figure 4.5).In any fluid flow, the fluid is contained within boundaries and it is assumed that a boundary layer exists where the fluid is in contact with the surface. This boundary layer is considered as stagnant and therefore has a velocity of zero. The velocity magnitude and direction of the moving fluid in contact with

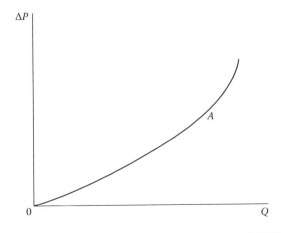

Figure 4.5 Relationship between pressure drop and flow rate for fluid flow in a straight, horizontal pipe.

this stationary layer will cause shear or frictional stress between the layers. The shear, which is possible between different elements of the fluid body, causes the pressure and shear stress to vary considerably from one point to another according to flow conditions.

Due to the complex nature of flow, precise mathematical analysis is limited to a few very specific cases. Flow problems are solved by experimentation or by making assumptions sufficient to simplify the mathematics to obtain a theoretical solution. The two approaches are not mutually exclusive, since the laws of mechanics are always valid and enable partially theoretical methods to be adopted in several important cases. It is important to determine experimentally how far the theoretical value will deviate from the "true" conditions as defined in the experiment.

4.4.2 Conservation of Mass

When fluid flows through a system at a constant flow rate where there are no accumulations or depletions of the fluid within the system, the conservation of mass within the system requires that

$$\dot{m}_1 = \dot{m}_2 = \dot{m}_3 = \cdots$$

where $\dot{m}_i$ = mass flow rate of fluid at point i (kilograms per second).

The mass flow rate at any point in the system is given by

$$\dot{m}_i = A_i V_i \rho_i$$

where A_i is the cross-sectional area (m²) of flow, V_i is the average velocity (m s⁻¹), and ρ_i is the fluid density (kg m⁻³)—all at point i.

$$A_1 V_1 \rho_1 = A_2 V_2 \rho_2 = A_3 V_3 \rho_3 = \cdots = \dot{m}$$

Since fluids are normally considered to be incompressible, the density can be assumed to be constant and the formula can be simplified to

$$A_1 V_1 = A_2 V_2 = A_3 V_3 = \cdots = Q$$

Example 4.5

Water is flowing in a 9 cm inside-diameter pipe at a velocity of 0.5 m s⁻¹ that enlarges to a 15 cm inside-diameter pipe. How will this change in diameter affect the velocity, the volumetric flow rate, and the mass flow rate in the larger section?

$$A_1V_1 = A_2V_2$$

$$V_2 = \frac{A_1}{A_2}V_1 = \left(\frac{D_1}{D_2}\right)^2 V_1$$

$$V_2 = \left(\frac{9 \text{ cm}}{15 \text{ cm}}\right)^2 (0.5 \text{ ms}^{-1}) = 0.18 \text{ ms}^{-1}$$

The long way to do this problem would be to convert the centimeters to meters and then half the diameter to get radius. You can then multiply the square of the radius with π to get the area in square meters. Doing this in a fraction is exactly the same as just using the ratio of diameter and, since the centimeters will cancel out, it also does not matter that you use centimeters instead of meters.

$$Q = A_1V_1 = \left[\pi\left(\frac{D_1}{2}\right)^2\right] V_1$$

$$Q = \pi\left(\frac{0.09 \text{ m}}{2}\right)^2 (0.5 \text{ ms}^{-1}) = 3.18 \times 10^{-4} \text{ m}^3\text{s}^{-1}$$

The volumetric flow rate will stay the same, as will the mass flow rate.

$$\dot{m} = Q\rho$$

$$\dot{m} = (3.18 \times 10^{-4} \text{ m}^3\text{s}^{-1})(1000 \text{ kgm}^{-3}) = 0.318 \text{ kgs}^{-1}$$

Remember that the quotients must be added when we multiply and $m^0 = 1$.

4.5 Types of Flow

4.5.1 Turbulent and Laminar Flow

In turbulent flow, the progression of the fluid particles is irregular and there is a seemingly haphazard interchange of position. Individual particles are subject to fluctuating transverse velocities so that the motion is eddying and sinuous rather than rectilinear. If dye is injected at a certain point, it will rapidly diffuse throughout the flow stream.

In laminar flow, all the fluid particles proceed along parallel paths and there is no transverse component of velocity. The orderly progression is such that each particle follows exactly the path of the particle preceding it without any deviation. Thus, a thin filament of dye will remain as such without diffusion. In a pipe, the ratio of the mean velocity V and the maximum velocity V_{max} is 0.5 with turbulent flow and 0.05 with laminar flow (see Figure 4.6).

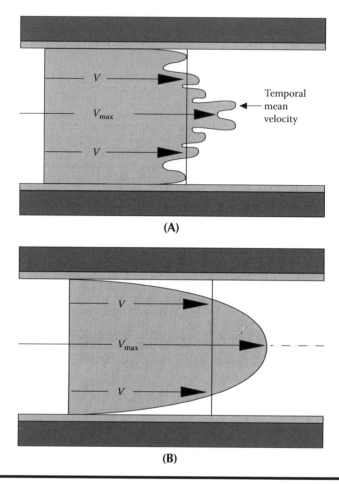

Figure 4.6 Characteristics of (A) turbulent and (B) streamline flow.

Laminar flow is associated with low velocities and viscous sluggish fluids. In pipeline and open channel hydraulics, the velocities are nearly always sufficiently high to ensure turbulent flow, although a thin laminar layer persists in proximity to a solid boundary. The laws of laminar flow are fully understood, and for simple boundary conditions, the velocity distribution can be analyzed mathematically. Due to its irregular pulsating nature, turbulent flow has defied rigorous mathematical treatment and, for the solution of practical problems, it is necessary to rely largely on empirical or semiempirical relationships.

4.5.1.1 Reynolds Number

Reynolds studied the flow behavior of liquids in pipelines at different velocities and, from the observations, he developed a mathematical relationship to define the conditions under which flow changes from laminar to turbulent. The fluid velocity at which the change occurs is called the critical velocity. Reynolds found that the average velocity of the fluid (V) (m/s), the diameter of the pipe (D) (m), the fluid density (ρ) (kg/m³), and the dynamic viscosity (μ) (Pa s) affect the type of flow in the following relationship:

$$Re = \frac{DV\rho}{\mu}$$

If we look at the units for the dimensions, we find

$$Re = \frac{m \times \dfrac{m}{s} \times \dfrac{kg}{m^3}}{Pas} = \frac{m \times \dfrac{m}{s} \times \dfrac{kg}{m^3}}{\dfrac{kg}{ms}} = m \times \frac{m}{s} \times \frac{kg}{m^3} \times \frac{ms}{kg} = 1$$

The Reynolds number (Re) is therefore dimensionless. It has been found that the flow will be laminar if the Re is less than 2,130 and it will be turbulent if Re is greater than 4,000. For values between 2,130 and 4,000 the characteristics of flow will depend upon the system and what the flow was prior to the change. When the velocity of a fluid that flows laminar is gradually increased, laminar flow can be maintained at rather high Re. In the same way, turbulent flow can persist at low Re once it has been established.

The Reynolds number can also be expressed in terms of mass flow rate rather than velocity. As stated earlier, the mass flow rate at any point in the system is given by

$$\dot{m} = AV\rho$$

where A is the cross-sectional area (m²) of flow, V is the average velocity (m s⁻¹), and ρ is the fluid density (kg m⁻³). Taking D as the average diameter, the cross-sectional area A can be expressed as

$$A = \frac{\pi D^2}{4}$$

This gives a Reynolds formula that is expressed in terms of mass flow rate:

$$Re = \frac{4\,\dot{m}}{\pi D \mu}$$

Example 4.6

Determine if the flow will be laminar or turbulent when a fluid with a density of 1.05 g/cm³ and a viscosity of 120 cP is pumped through a 3 cm inside-diameter steel pipe at 100 L/min (volumetric flow rate Q).

$$D = 3 \text{ cm} \times \frac{m}{100 \text{ m}} = 0.03 \text{ m}$$

$$Q = \frac{100 \text{ dm}^3}{\text{min}} \times \frac{m^3}{1000 \text{ dm}^3} \times \frac{\text{min}}{60 \text{ s}} = 1.667 \times 10^{-3} \text{ m}^3 \text{ s}^{-1}$$

$$V = \frac{Q}{\text{area}} = \frac{1.667 \times 10^{-3} \text{ m}^3 \text{ s}^{-1}}{\frac{\pi}{4} \times (0.03 \text{ m})^2} = \frac{1.667 \times 10^{-3} \text{ m}^3 \text{ s}^{-1}}{7.069 \times 10^{-4} \text{ m}^2} = 2.358 \text{ m/s}$$

$$\rho = \frac{1.05 \text{ g}}{\text{cm}^3} \times \frac{\text{kg}}{1000 \text{ g}} \times \frac{1 \times 10^6 \text{ cm}^3}{m^3} = 1050 \text{ kg/m}^3$$

$$\mu = 120 \text{ cp} \times \frac{0.001 \text{ Pas}}{\text{cp}} = 0.12 \text{ Pas}$$

$$\text{Re} = \frac{DV\rho}{\mu} = \frac{0.03 \times 2.358 \times 1050}{0.12} = 618.98$$

The flow will be laminar since Re < 2100.

4.5.2 Rotational and Nonrotational Flow

Flow is said to be rotational if each fluid particle has an angular velocity about its own mass center. Flow in a circular path is depicted with the velocity directly proportional to the radius. The two axes of the particle rotate in the same direction so that the flow is again rotational.

For the flow to be nonrotational, the velocity distribution adjacent to the straight boundary must be uniform. In the case of flow in a circular path, it may be shown that nonrotational flow will only pertain if the velocity is inversely proportional to the radius. The two axes rotate in opposite directions, so there is a compensating effect producing an average orientation of the axes, which is unchanged from the initial state.

Because all fluids possess viscosity, the flow of a real fluid is never truly nonrotational, and laminar flow is, of course, highly rotational. Thus, nonrotational flow is a hypothetical condition that would be of academic interest only were it not for the fact that in many instances of turbulent flow the rotational characteristics are so insignificant that they may be neglected. This is convenient because it is possible to analyze nonrotational flow by means of the mathematical concepts of classical hydrodynamics referred to earlier.

4.5.3 Steady and Unsteady Flow

Flow is said to be steady when the conditions at any point are constant with respect to time. A strict interpretation of this definition would lead to the conclusion that turbulent flow was never truly steady. However, for the present purpose, it is convenient to regard the general fluid motion as the criterion and the erratic fluctuations associated with the turbulence as only a secondary influence. An obvious example of steady flow is a constant discharge in a conduit or open channel.

As a corollary, it follows that the flow is unsteady when conditions vary with respect to time. An example of unsteady flow is a varying discharge in a conduit or open channel; this is usually a transient phenomenon being successive to, or followed by, a steady discharge. Other familiar examples of a periodic nature are wave motion and the cyclic movement of large bodies of water in tidal flow.

Most of the practical problems in hydraulic engineering are concerned with steady flow. This is fortunate, since the time variable in unsteady flow considerably complicates the analysis.

4.5.4 Uniform and Nonuniform Flow

Flow is uniform when there is no variation in the magnitude and direction of the velocity vector from one point to another along the path of flow. For compliance with this definition, both the area of flow and the velocity must be the same at every cross section. Nonuniform flow occurs when the velocity vector varies with the location; a typical example is flow between converging or diverging boundaries.

Both of these alternative conditions of flow are common in open channel hydraulics. Since uniform flow is always approached asymptotically, it is an ideal state that is only approximated and never actually attained. It should be noted that the conditions relate to space rather than time and therefore, in cases of enclosed flow (i.e., pipes under pressure), they are quite independent of the steady or unsteady nature of the flow.

4.6 Conservation of Energy in Liquids

Since energy is not created or destroyed within the flow system, the energy must balance throughout the system. When we consider potential energy (PE), kinetic energy (KE), added energy (AE), and lost energy (LE), the following will hold:

$$PE_1 + KE_1 + AE - LE = PE_2 + KE_2$$

The potential energy, head of liquid, at any point within the system has two components: namely, elevation energy (potential head) and pressure energy (pressure head). It is convenient to describe the position, pressure, and velocity of a liquid in terms of the equivalent static heads.

4.6.1 Potential Head

If a quantity of fluid is situated at an instant at a height (h) vertically above some arbitrarily chosen datum, the potential head (h) is given in meters above the datum.

4.6.2 Pressure Head

Consider a column of liquid of height h, cross-sectional area a, and density ρ:

$$\text{Volume of the column} = ah$$

$$\text{Mass of the column} = ah\rho$$

$$\text{Force due to mass of the column} = ah\rho g$$

Then the pressure on the base of the column due to its force

$$P = \frac{ah\rho}{a} = h\rho g$$

$$h = \frac{P}{\rho g}$$

where h is the pressure head.

4.6.3 Velocity Head

When a mass m (kilograms) of fluid at rest (zero initial velocity) falls freely through a height h, then loss in potential energy = gain in kinetic energy and the velocity head h is given by

$$mgh = \frac{mV^2}{2}$$

$$h = \frac{mV^2}{2\,mg} = \frac{V^2}{2\,g}$$

Adding these results gives

$$\text{Total head of a liquid} = h + \frac{P}{rg} + \frac{V^2}{2g} \ (\text{meters})$$

4.6.4 Energy Changes and Flow

When fluid flows through a system, there are frictional losses and changes in kinetic energy, potential energy, and pressure energy. Bernoulli's equation can be used to calculate the energy balance when liquid flows through pipes, the effects of the fittings within the system, or the pressure developed by a pump:

$$\frac{P_1}{\rho_1} + \frac{V_1^2}{2} + h_1 g = \frac{P_2}{\rho_2} + \frac{V_2^2}{2} + h_2 g$$

P (Pa) is the pressure, ρ (kg m^{-3}) is the fluid density, g (9.81 ms^{-2}) is the acceleration due to gravity, V (ms^{-1}) is the velocity of the fluid, and h (m) is the height above a common datum. The terms in Bernoulli's equation for fluids are work or energy per volumes (J/m^3).

When we add energy and take frictional losses into account, this formula can be written as

$$E_P - E_F + \frac{P_1}{\rho_1} + \frac{V_1^2}{2} + h_1 g = \frac{P_2}{\rho_2} + \frac{V_2^2}{2} + h_2 g$$

E_P is the energy added with a pump and E_F is energy lost through friction.

The principles of Bernoulli's equation can be applied to many situations. When flow is horizontal, $h_1 = h_2$; then,

$$P + \frac{1}{2}\rho V^2 = \text{constant}$$

If the flow is vertical down (+) or up (–), then the difference in height must be considered.

This indicates that pressure decreases if the fluid velocity increases and vice versa. When liquid thus flows through part of a piping system that has a smaller diameter than the rest, the velocity of the liquid will increase and the pressure will drop.

Example 4.7

A frictionless, incompressible fluid with a density of 1100 kg/m^3 flows in a pipeline. At a point where the pipe diameter is 15 cm, the fluid velocity is 2 m/s and the pressure is 300 kPa. Determine the pressure at a point 10 m downstream where the diameter is 5 cm if the pipe is horizontal.

Since $h_1 = h_2$, the two terms $h_1 g$ and $h_2 g$ will cancel out:

$$A_1 V_1 = A_2 V_2$$

$$V_2 = \frac{A_1 V_1}{A_2} = V_1 \left(\frac{D_1}{D_2} \right)^2 = 2 \text{ m/s} \left(\frac{15}{5} \right)^2 = 18 \text{ m/s}$$

$$P_2 = \frac{P_1 \rho_2}{\rho_1} + \frac{\rho_2 (v_1^2 - v_2^2)}{2}$$

$$P_2 = \frac{300 \text{ kPa} \times 1100 \text{ kg/m}^3}{1100 \text{ kg/m}^3} + \frac{1100 \text{ kg/m}^3 [(2 \text{ m/s})^2 - (18 \text{ m/s})^2]}{2}$$

$$P_2 = 300 \text{ kPa} + [550 \text{ kg/m}^3 \times (4 - 324) \text{m}^2/\text{s}^2]$$

$$P_2 = 300 \text{ kPa} + [-176000 \text{ N/m}^2] = 300 - 176 \text{ kPa}$$

$$P_2 = 124 \text{ kPa}$$

Example 4.8

For the same data, what would the pressure be if the flow were vertical downward?

$$P_2 = \frac{P_1 \rho_2}{\rho_1} + \frac{\rho_2 (V_1^2 - V_2^2)}{2} + \rho_1 (h_1 - h_2) g$$

$$P_2 = (300 \text{ kPa} - 176 \text{ kPa}) + 1100 \text{ kg/m}^3 (10 \text{ m} - 0 \text{ m}) 9.81 \text{ m/s}^2$$

$$P_2 = 231.91 \text{ kPa}$$

4.6.5 Frictional Losses

Fluid flowing along a straight smooth pipeline will suffer losses in energy caused by viscosity (resistance to flow). In most pipelines the flow will be subjected to change of direction, valves, changes in pipe diameter, flow meters, and many other devices that will cause energy losses.

4.6.5.1 Frictional Losses in Straight Pipes

Bernoulli's equation can be applied between two points in a straight pipe to give

$$F = \frac{P_1 - P_2}{\rho g} = \frac{P}{\rho g}$$

When the flow is horizontal and there is no change in pipe diameter, there is no change in velocity and kinetic or potential energy. The frictional loss F is related to the pressure drop per unit length of the pipe. When flow is laminar, Re < 2100, the pressure drop per unit length $\Delta P/L$ is related to the pipe diameter D, density ρ, and average velocity V through the Fanning friction factor.

$$f = \frac{PD}{2\rho V^2 L}$$

At this, the flow can be written in terms of volumetric flow, Q:

$$Q = \frac{P\pi D^4}{128 L \mu}$$

The Hagen-Poiseuille equation can be derived from it to give a friction factor:

$$f = \frac{16}{Re}$$

Using the frictional loss equation gives

$$F = \frac{P}{\rho g} = \frac{128 L \mu Q}{\pi D^4 \rho g}$$

where L is the pipe length in meters. Under turbulent flow conditions in smooth pipe, the situation is more complex and the Blasius equation is used to calculate the friction factors for Reynolds numbers 4,000 < Re < 100,000.

$$f = \frac{0.079}{Re^{0.25}}$$

$$f = 0.079 Re^{-0.25}$$

To obtain the frictional losses in a pipe when flow is laminar or turbulent, the friction factor must be determined and the values placed in the Fanning equation:

$$F = \frac{P}{\rho} = 2f \frac{LV^2}{Dg}$$

Example 4.9

Water is flowing through a 20 m smooth pipe with an inside diameter of 25 mm. Calculate the frictional loss when the average velocity (V) is 2 m/s.

$$L = 20 \text{ m}$$

$$v = 2 \text{ m/s}$$

$$\rho = 1000 \text{ kg/m}^3$$

$$\mu = 10^{-3} \text{ Nsm}^{-2}$$

$$D = 25 \text{ mm} \times \frac{1 \text{ m}}{1000 \text{ mm}} = 0.025 \text{ m}$$

$$Re = \frac{DV\rho}{\mu} = \frac{0.025 \times 2 \times 1000}{10^{-3}} = 50000 = 5 \times 10^4$$

The flow is turbulent.

$$f = 0.079 \, Re^{-0.25} = 0.079 \times (5 \times 10^4)^{-0.25} = 5.28 \times 10^{-3}$$

$$F = 2f \frac{LV^2}{Dg} = 2 \times 5.28 \times 10^{-3} \left(\frac{20 \text{ m} \times (2 \text{ m/s})^2}{0.025 \text{ m} \times 9.81 \text{ m/s}^2} \right)$$

$$F = 10.56 \times 10^{-3} \left(\frac{80 \text{ m}^3 \text{ s}^2}{0.2453 \text{ m}^2/\text{s}^2} \right) = 3.44 \text{ m}$$

Changing the pipe diameter will cause different frictional losses. With lower frictional losses, the pumping costs will be reduced but the pipe cost will increase. By doing the calculations, one will be able to compare the one-time installation cost with the continual running cost.

4.6.5.2 Fittings in Pipes

Nearly all fittings and irregularities in pipes cause energy loss. Table 4.2 gives some fittings with the equivalent length of uniform pipe that would cause the same loss. The friction loss term in meters is expressed as

$$F + K \frac{V^2}{2g}$$

Once all the fittings are converted into length of pipe, the length is added to L in the Fanning equation and the friction caused by pipe fittings can be calculated.

Table 4.1 Surface Tension of Some Liquids

Liquid	Surface Tension (mN m⁻¹)
Water	72.75
Ethyl alcohol	22.75
Methyl alcohol	22.65
Carbon tetrachloride	26.95
Skim milk (0.04% fat)	51.0
Glycerol	63.4
Mercury	435.5

Example 4.10

There are four 90° elbows, one ball valve, and one nonreturn valve in a 6 m, 25 mm internal diameter pipeline. What length should be used to calculate the friction losses?

$$6 + 4 \times 0.84 + 8.23 + 1.58 = 19.17 \text{ m}$$

4.6.6 Mechanical Energy Requirement

The mechanical energy requirement can be calculated by using Bernoulli's equation. Some assumptions will make using it much less complicated:

$$E_P - E_F + \frac{P_1}{\rho_1} + \frac{V_1^2}{2} + h_1 g = \frac{P_2}{\rho_2} + \frac{V_2^2}{2} + h_2 g$$

Rearranging the equation to get the energy supplied by the pump (E_P) on the left and all else on the right yields

$$E_P = \frac{P_2}{\rho_2} - \frac{P_1}{\rho_1} + \frac{V_2^2}{2} - \frac{V_1^2}{2} + h_2 g - h_1 g + E_F$$

One can frequently remove some terms. If the velocity in the system is the same at both points, then the difference in velocity will be zero. If the difference in pressure is very small and the difference in density is also very small, the difference between the two pressure and density ratios will tend toward zero and can be ignored. That will leave

$$E_P = (h_2 - h_1)g + E_F$$

This formula requires very special conditions and assumptions.

Example 4.11

A fluid with a viscosity of 1.2 cP (0.0012 Pa s) is pumped from one tank to another where the free surfaces of the liquids in the tanks differ 12 m. If the total flow resistance in the system is 56 m²/s², calculate the pump energy required to pump the liquid at a rate of 30 m/min (0.5 m/s).

Since $h_1 = 0$, $V_1 = 0$, and $P_1 = P_2$ = atm. on the surfaces and since the density does not change, the equation becomes

$$E_P = \frac{V_2^2}{2} + h_2 g + E_F$$

$$E_P = \frac{(0.5 \text{ m/s})^2}{2} + 12 \text{ m} \times 9.81 \text{ m/s}^2 + 56 \text{ m}^2/\text{s}^2 = 67.9 \text{ m}^2/\text{s}^2 = 67.9 \text{ J/kg}$$

Problems

1. Express the pressure in SI units at the base of a column of fluid 8.325 in. high when the acceleration due to gravity is 32.2 ft/s² and the fluid density is 1.013 g/cm.

$$P = h\rho g$$

2. Calculate the power available (W) in a fluid that flows down the raceway of a reservoir at a rate of 525 lb_m/min from a height of 12.3 ft. The potential energy is given by PE = mgh (m = mass, g = acceleration due to gravity [32.2 ft/s²], and h = height).

$$h = 8.325 \text{ in} = 8.325 \text{ in} \times \frac{0.0254 \text{ m}}{\text{in}} = 0.211 \text{ m}$$

$$\rho = 1.013 \text{ g/cm}^3 = \frac{1.013 \text{ g}}{\text{cm}^3} \times \frac{1 \text{ kg}}{1000 \text{ g}} \times \frac{1 \times 10^6 \text{ cm}^3}{\text{m}^3} = 1013 \text{ kg/m}^3$$

$$g = 32.2 \text{ ft/s}^2 = \frac{32.2 \text{ ft}}{\text{s}^2} \times \frac{1 \text{ m}}{3.28084 \text{ ft}} = 9.79 \text{ m/s}^2$$

$$P = h\rho g$$

Just substitute the answers and do the final calculation. It is important to remember that W = J/s and that is why we use J/s in the initial calculation.

$$PE = \frac{Ykg}{s} \times \frac{9.81 \text{ m}}{s^2} \times Xm = AJ/s = AW$$

$$J = \text{kgm}^2/s^2$$

$$J/s = W$$

3. In a heat transfer system where the product can be damaged through local overheating, you have to ensure turbulent flow. If the product has a density of 1.3 g/mL, a viscosity of 20.2 cP, and flows at 25 L/min, what is the maximum diameter pipe that can be used to ensure turbulent flow?

$$\text{Re} = \frac{4 \text{ m}}{\pi D \mu}$$

$$D = \frac{4 \text{ m}}{\text{Re} \pi \mu}$$

For flow to be turbulent, the Reynolds number must be greater than 4,000.

$$\rho = \frac{1.3 \text{ g}}{\text{ml}} \times \frac{1 \text{ kg}}{1000 \text{ g}} \times \frac{1{,}000{,}000 \text{ ml}}{\text{m}^3} = A \text{ kg/m}^3$$

$$m = \frac{25 \text{ L}}{\text{min}} \times \frac{1 \text{ m}^3}{1000 \text{ L}} \times \frac{1 \text{ min}}{60 \text{ s}} \times \frac{A \text{ kg}}{\text{m}^3} = B \text{ kg/s}$$

$$\mu = 20.2 \text{ cP} \times \frac{0.001 \text{ Pas}}{\text{cP}} = C \text{ Pas}$$

$$D = \frac{4 \text{ m}}{\text{Re} \pi \mu}$$

$$D = \frac{4 \times B \text{ kg/s}}{4000 \times \pi \times C \text{ Pas}}$$

$$D = E \times 10^{-3} \text{ m} = E \times 1000 \text{ mm}$$

Now we can check the answer by doing the reverse and calculate the Reynolds number:

$$A = \frac{\pi D^2}{4} = F \times 10^{-5} \text{ m}^2$$

$$Q = \frac{25 \text{ L}}{\text{min}} \times \frac{\text{m}^3}{1000 \text{ L}} \times \frac{\text{min}}{60 \text{ s}} = G \times 10^{-4} \text{ m}^3/\text{s}$$

$$V = \frac{Q}{A} = \frac{G \times 10^{-4} \text{ m}^3/\text{s}}{F \times 10^{-5} \text{ m}^2} = H \text{ m/s}$$

$$\text{Re} = \frac{DV\rho}{\mu} = \frac{E \times 10^{-3} \text{ m} \times H \text{ m/s} \times A \text{ kg/m}^3}{C \text{ Pas}}$$

$$\text{Re} = 3998.4$$

This is close enough to the minimum of 4,000.

4. Water at 20°C is flowing a rate of 20 L/minute through a 5 cm internal diameter pipe. The viscosity of water at this temperature is 0.98 cP. Calculate the Reynolds number.

5. If the pipe in problem 4 narrows to 2 cm, what will the Reynolds number be? If we assume frictionless flow, calculate the difference of pressure between the two sections of pipe.

Chapter 5

Pumps

5.1 Introduction

Transfer of liquids and gases is a basic operation in many manufacturing plants and is frequently taken for granted. We accept water flowing from a faucet as normal, the way it should be. We seldom think about the process to pump water from a well, through various treatments, and into an elevated reservoir from where it flows into our homes.

Pipelines and pumps are efficient means of transferring liquids and gases from one operation to another. For robotic and automated processes, operations frequently involve hydraulic fluid transfer and control. Many components in high-speed machinery require constant lubrication by special pumps to keep them working.

5.2 General Principles

No matter what needs to be pumped, the pump selected should be able to do the work at minimum cost and maximum efficiency and reliability. The pumping of wide varieties of fluids through the different pieces of equipment adds to the complexity of the pumping process.

A pump is a machine that transfers mechanical energy to a fluid. The term "pump" is reserved for machines that handle incompressible fluids. The machines that handle compressible fluids are called compressors. In a pumping operation, the pressure of the fluid will increase and cause it to flow downstream in a direction of lower pressure. On the intake side, we can also think of a pump as a classic suction system, where the pressure is reduced to a level at which fluid flows toward the pump.

To transport fluid, it is essential that enough energy be added to overcome frictional losses. Many factors are considered when the size and type of pump are selected:

1. The pressure required (pressure head)
2. Volumetric flow rate (velocity head)
3. Properties of the fluid handled
   a. Density
   b. Viscosity
   c. Oxidation sensitivity
   d. Abrasiveness
   e. Flow properties (Newtonian or non-Newtonian)
   f. Foaming
   g. Shear damage
4. Temperature of the fluid and the vapor pressure
5. Operational considerations such as intermittent or continued use

The many types of pumps can be classified in two broad categories: positive displacement pumps and dynamic pumps. Positive displacement pumps will normally produce a low, consistent flow rate at sometimes very high pressures. The volume of the fluid delivered is independent of the discharge pressure. Dynamic pumps usually give high flow rates at relatively low pressures and the discharge pressure directly affects the volume of delivered fluid.

5.3 Pump Efficiency

The mechanical efficiency of a pump can be calculated from the ratio of power output (p_o) to power input (p_i):

$$e_m = \frac{p_o}{p_i}$$

The power output (p_o) is directly related to the product of work done (W) to the fluid (J/N), the volumetric flow rate (Q) of the fluid (m^3/s), the density (ρ) of the fluid (kg/m^3) and acceleration due to gravity (g) (m/s^2).

$$p_o = WQ\rho g$$

Two shortcut formulas are given by Perry[*]:

---

[*] Perry, R. H. 1997. *Perry's Chemical Engineers' Handbook*, 7th ed., ed. D. W. Green. New York: McGraw–Hill.

$$kW = \frac{hQ\rho}{3.670 \times 10^5}$$

where kW is the pump power output (kilowatts), h is the total dynamic head in water column (Nm/kg), Q is the capacity (m³/h) and ρ is the density (kg/m³). When the total dynamic head is expressed in pascals, then the following formula should be used:

$$kW = \frac{hQ}{3.599 \times 10^6}$$

Example 5.1

A pump delivers 13 L of water per second with a total head of 12 m. What is the power output of the pump?

$$p_o = WQ\rho g$$

$$p_o = (12 \text{ m}) \left(\frac{13 \text{ dm}^3}{s} \times \frac{m^3}{1000 \text{ dm}^3} \right) \left(\frac{1000 \text{ kg}}{m^3} \right) \left(\frac{9.81 \text{ m}}{s^2} \right)$$

$$p_o = \frac{1530360 \text{ mdm}^3 \text{ m}^3 \text{ kgm}}{1000 \text{ sdm}^3 \text{ m}^3 \text{ s}^2} = 1530.36 \text{ kgm}^2/s^3$$

$$p_o = \frac{1530.36 \text{ kgm}^2}{s^3} \times \frac{1 \text{ N}}{1 \text{ kgms}^{-2}} = 1530.36 \text{ Nm/s}$$

$$p_o = \frac{1530.36 \text{ Nm}}{s} \times \frac{1 \text{ J}}{1 \text{ Nm}} = \frac{1530.36 \text{ J}}{s} \times \frac{1 \text{ W}}{1 \text{ Js}^{-1}} = 1530.36 \text{ W}$$

What is the mechanical efficiency of the pump if the power input is 3 hp where 1 hp is equal to 745.7 W?

$$e_m = \frac{p_o}{p_i} = \frac{1530.36 \text{ W}}{3 \text{ hp}} \times \frac{1 \text{ hp}}{745.7 \text{ W}} = 0.684 = 68.4\%$$

5.4 Centrifugal Pumps

Centrifugal pumps are designed to handle liquids that, for all practical purposes, are incompressible. They will not function as normal pumps when handling vapors or gases. When vapors or gases are present in the pump, they interfere with its normal operation and have a marked effect on its characteristics.

By definition, centrifugal pumps depend on centrifugal force for their operation. When an object moves around in a circle at a constant tip speed, it is constantly changing its velocity due to the change in direction. When water is stirred, it has a spiral flow toward the axis of rotation. The velocity of rotation is known as "velocity of whirl." For a free vortex, the product of the velocity of whirl (v) and the radius (r) will be constant.

$$V_1 r_1 = V_2 r_2$$

This means that the velocity continuously decreases at points further away from the axis of rotation.

In a forced vortex, external force is applied to produce and maintain the vortex. A forced vortex is only possible in real fluids. The angular velocity (ω) is constant throughout the vessel:

$$V = \omega r$$

At a larger radius, the velocity V will increase. Since the angular velocity (ω) is constant throughout the vessel, it follows that velocity (V) will increase with the increase of circumferential distance traveled. Velocity head was previously defined as:

$$h = \frac{V^2}{2\,g}$$

Substituting for V, we get:

$$h = \frac{\omega^2 r^2}{2\,g}$$

The velocity head is directly proportional to the radius and angular velocity, the size, and rotational speed. The two main groups of centrifugal pumps are single-stage and multistage pumps. The pressure (head) that can be generated by one impeller (stage) can vary from 10 to 100 m. Some chemical pumps can generate a head of up to 200 m with a single impeller. Single-stage pumps that can produce high pressure normally have an exponential increase in cost.

In a particular design, the pressure generated by a single impeller might not be sufficient and a series of impellers are used. These multistage pumps have the same basic design requirement for their main components, except that the discharge from one impeller is diverted to the intake for the next impeller. The impellers may be mounted on the same shaft or, in some designs, they are mounted on different shafts. Multistage centrifugal pumps are used when a high head is required or when the required volume capacity is large. When a high head is required, the impellers are normally mounted on the same shaft. To increase volumetric flow, the impellers are normally mounted on different shafts. It is cheaper to use a multistage pump to produce a high head than to use a single-stage pump to create the same head.

5.4.1 Operation

The impeller is designed to set up a forced vortex with low pressure in the center and high pressure at the periphery. The tendency of fluid is to flow into the center and out at the periphery.

The efficiency of a centrifugal pump is dependent upon the form of the impeller. The number of vanes varies from one to eight or more depending upon the conditions of use. Open vane impellers are used for pumping fluids containing small amounts of solids. Semi-open impellers can handle fluids containing sediment and other foreign matter. Enclosed impellers are nonclogging and do not require close running clearances for optimal performance.

Fluid enters the pump at the center of the impeller, from where it is moved by the revolving blades of the impeller. The rotation generates centrifugal force, resulting in a pressure at the outer diameter of the impeller. When flow takes place, the fluid passes from the impeller at considerable velocity and pressure to a passageway in the casing that will gradually expand to the discharge section (see Figure 5.1).

The casing has a volute shape (a spiral form that constantly recedes from the center). The impeller discharges the fluid into this passageway. The volute collects the water and directs it to the discharge opening. The volute is proportioned in such a way that it will gradually change the velocity around the circumference. The velocity head has to be changed into a pressure head (see Figure 5.2).

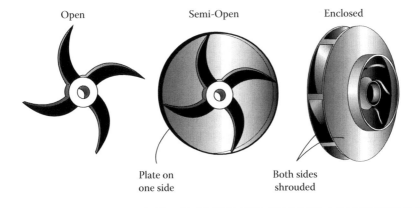

Figure 5.1   Impeller types.

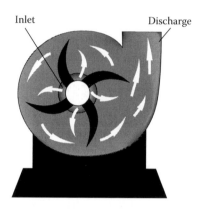

Figure 5.2   Section of a volute centrifugal pump.

The pump will only operate if the casing is filled with fluid. If vapor or gases are present in the pump, they will interfere with the normal operation and they influence the pump characteristics.

Any amount of vapor or gas present in the pump will cause the gas or vapor to be mixed with the fluid, thereby causing enhanced oxidation. Vigorous churning can sometimes result in the formation of unwanted but stable foam. (Ketchup is a good example.) The pumping performance will also decrease since the gas will decrease the fluid flow rate.

Cavitation (formation and collapse of vapor cavities in a flowing liquid) results in noise, vibration, metal erosion, and bearing failure. When the local pressure at a specific temperature inside a system is reduced to be equal to the vapor pressure of the fluid at that temperature, the fluid will boil and form vapor (see Table 5.1). As discussed earlier, the pressure will decrease when the flow increases at a place where pipe diameter decreases and vapor bubbles will form. As soon as the fluid flows into an area of higher pressure, the bubbles will collapse. This will cause vibration and excessive wear to the pipes in the area.

In a centrifugal pump, the area with the lowest pressure is the entrance to the impeller vanes. This is the place where vaporization will take place. Since the formation of vapor will interfere with performance, it is important that the formation

Table 5.1   Boiling Points of Fluids (°C) at Various Pressures

Liquid	101.3 kPa	31.22 kPa.	20.05 kPa	6.51 kPa	1.43 kPa
Water	100	70	60.3	38.4	14.87
Alcohol	78.2	51	42	24	–3.1
Ether	35	5	–5	–25	–55
Glycerin	290	252.5	240	215	177.5

of vapor be avoided. Keeping the pressure at this point above the vapor pressure of the liquid will prevent the formation of vapor.

The suction (low-pressure) part of the system starts from the free surface of the liquid in a tank and ends at the entrance to the impeller vanes. Centrifugal pumps cannot impart energy to a liquid until it reaches the impeller vanes. The energy to overcome frictional and other losses in the suction system in order to bring the liquid to the vanes must come from a source external to the pump such as atmospheric pressure or head.

The total dynamic head of a pump is the total discharge head minus the total suction head. The total suction head is the difference between the readings on a gauge at the centerline of the suction flange on a pump and a gauge on the discharge of the pump at the centerline. The static suction head is the vertical distance measured from the free surface of the liquid source to the pump centerline plus the absolute pressure at the liquid surface. If the water level is higher than the centerline of the pump on the inlet side, the pump is operating under inlet head.

Please remember that nothing can "suck"[*]; flow is initiated when a pressure that is higher than the local pressure pushes the fluid toward the low pressure. In certain applications this push can be over a very long distance.

5.4.1.1 Net Positive Suction Head

To prevent vaporization of the fluid in the low-pressure part of the pump, the net positive suction head (NPSH) for the application is established. The NPSH values depend upon the fluid to be pumped as well its physical properties, such as the boiling point. In the case of mixtures, it is the boiling point of the compound with the lowest boiling point that will be critical.

5.5 Suction Systems and Sump Design

Vapors or gases inside a pump will cause serious problems. Water and other liquids normally contain dissolved air and other gases (CO_2, N, etc.) unless it has been deaerated. The amount of dissolved gas depends on the pressure and temperature of the liquid. When pressure is reduced, the dissolved gases will evaporate in the form of bubbles, which, due to expansion at low pressures, will occupy relatively large volumes. It is important to remember that air and gas are normally more soluble in liquids at low temperatures than in liquids at higher temperatures.

It is also possible that air enters the suction pipeline through poor design. In most cases, a simple change in pipeline configuration will stop the problem. When entrained air enters the system or when air comes out of solution due to low pressure, the air should be removed through the pump as soon as it forms. This will

---

[*] Remember "negative" pressure!

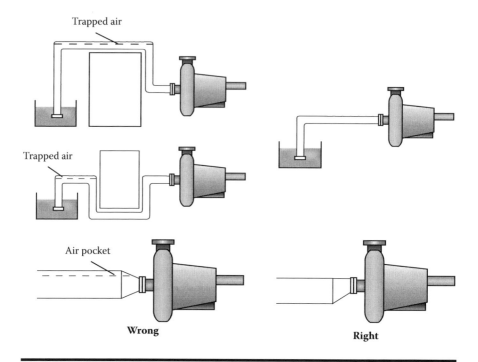

Figure 5.3   Suction pipe arrangements that will trap air.

prevent large amounts of air from entering the pump and seriously affecting the pump's performance (Figure 5.3).

The selection of the suction pipe size and design is critical for the good performance of the system. Most of the problems with centrifugal pumps are attributable to poor inlet system design. The inlet pipe should not be smaller than the inlet opening of the pump. The inlet pipe should be as short and as direct as possible. Where long inlet lines cannot be avoided, the diameter of the pipe should be increased. Air pockets or high points in the inlet line will cause trouble. Inlet piping can be horizontal but it is preferable to have a continuous slight rise toward the pump. Eccentric reducers should always be used. Pipe fittings (like elbows or tees) that are located close to the pump suction flange will result in uneven flow patterns.

The design of the supply or sump tank should receive good attention. Many problems can be avoided if one takes a few precautions. Large suction pipes should be submerged about four times their diameter while small pipes require 0.5 to 1 m submergence. The amount of entrained air that can enter the suction line should be kept to a minimum. When liquid is discharged into the sump, steps must be taken to place the pump inlet so that the air will not cause a problem (Figure 5.4).

The baffles in the supply tank will solve the problem by diverting any entrained air away from the suction pipe (Figure 5.5).

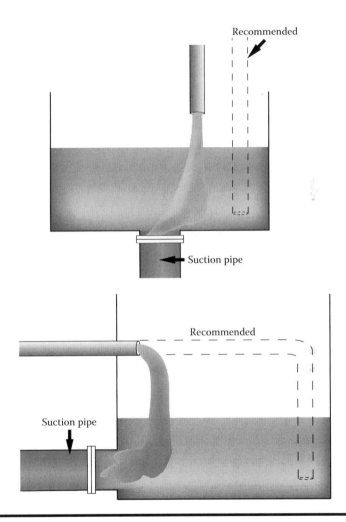

Figure 5.4 Sump design that can cause problems due to entrained air entering the suction pipe.

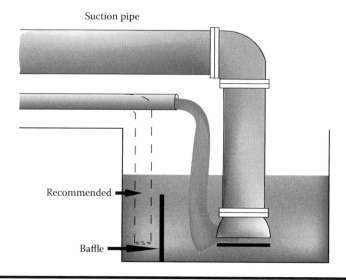

Figure 5.5   Rectifying the entrained air problem by inserting a baffle.

Insufficient submergence of the pump suction line can cause a vortex, which is a swirling funnel of air from the surface directly into the pump suction line.

The flat baffle must extend in all directions far enough from the suction pipe entry to avoid a vortex (Figure 5.6). The entry into the suction pipe should also receive attention. Flanging the pipe inlet to a bell mouth will prevent the vortex to some extent. It is also possible to design the suction inlet such that the formation of a vortex is reduced (Figure 5.7).

A strainer is frequently attached to the end of the inlet pipe to prevent foreign material from lodging in the impeller. The strainer should be at least three to four times the area of the inlet pipe. When pumps are used intermittently, a foot valve is essential to maintain the prime. The submergence of the suction pipe must also be carefully considered. The amount of submergence required depends on the size and capacity of individual pumps as well as on the sump design.

5.6 Positive Displacement Pumps

Positive displacement pumps operate by "sucking" a specific volume of fluid into the pump and then forcing this volume out at the discharge side. It will deliver a specified volume of fluid irrespective of the head. It is for this reason that a pressure safety valve close to the pump outlet is normally part of the piping system. These pumps are self-priming. In positive displacement pumps, the fluid is directly displaced and the capacity is independent of pressure and directly proportional to speed.

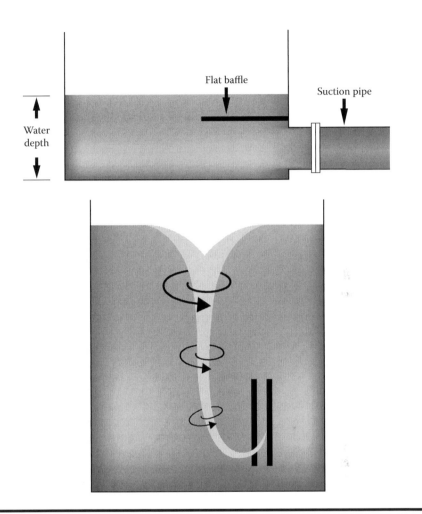

Figure 5.6   Rectifying vortex formation with baffles.

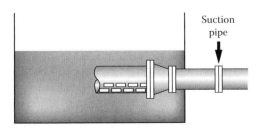

Figure 5.7   End fitting on a suction pipe to reduce vortex formation.

Positive displacement pumps have the following characteristics that are essential requirements in many process applications:

1. **Capacity is proportional to speed.** The capacity of these pumps is proportional to pump speed and is independent of pressure head. Centrifugal pumps deliver head; positive displacement pumps deliver pressure. When variable speed drives are installed on these pumps, the output can be very easily adjusted. The positive displacement pumps deliver a steady flow rate but they are not always truly positive. There is a certain amount of internal leakage or slip that occurs at very high pressures so that the flow rate will decrease slightly when the outlet pressure increases. This is independent of pump speed. When the pressure is small, no slip occurs, and the pump will operate at 100 percent volumetric efficiency. Slip can also be caused by factors such as fluid viscosity and clearances between elements.

2. **Head is independent of pump speed.** Varying pump speeds will have a negligible effect on the capacity of the pump. It will deliver the same volume per rotation at whatever head or speed. Positive displacement pumps cannot operate against closed discharge lines without damage to the pump, the pipeline, or other process equipment. A bypass relief valve is essential in the discharge line.

3. **Power requirements are proportional to head.** At constant speed, the power requirement will be directly proportional with the developed head. If speed remains constant, power requirements are determined by the capacity pumped dynamic head, and efficiency.

4. **Self-priming is a design feature.** Most rotary positive displacement pumps are self-priming. There are, however, certain aspects, such as damage to seals when the pump runs dry, that must be considered.

5. **Fluids containing entrained gases or vapors can be handled.** Rotary pumps move predetermined units of fluid with every rotation and they can pump fluids containing gases or vapors without loss of prime. Remember that there will be a loss of volume since air is moved and at the high pressure it will dissolve. This can cause many problems when the fluid exits the system and gets into normal pressure environments.

6. **Pumps can be reversed.** Progressive cavity type pumps can pump in either direction with equal efficiency, simply by changing the direction of rotation.

7. **Suction capability is proportional to speed.** Suction capability of the pump increases with increasing pump speed. At low pump speeds, suction is low, and volatile liquids can be handled.

8. **Accurate proportioning can be accomplished.** Reciprocating piston pumps can be used for accurate metering or proportioning applications.

9. **Power requirements are variable.** The power requirements of rotary pumps depend upon displacement capacity and total differential pressure across the pump. They are independent of the characteristics of the fluid that is pumped.

5.6.1 Rotary Pumps

Pumps of this type include helical screw pumps, lobe pumps, gear pumps, and peristaltic pumps. In rotary pumps, relative movement between rotating elements and the stationary element of the pump causes the pumping action. The operation is different from reciprocating pumps, where valves and a piston are integral to the pump. They also differ from centrifugal pumps where high velocity is turned into pressure. Rotary pumps are designed so that a continuous seal is maintained between inlet and outlet ports by the action and position of the pumping elements and close running clearances of the pump. Rotary pumps therefore do not require valve arrangements like certain reciprocating pumps.

5.6.1.1 External Gear and Lobe Pumps

5.6.1.1.1 External Gear Pumps

In the external gear pump, a pair of meshing gears of the same size rotates in a close-fitting casing. One or both gears can be coupled to a drive shaft through which the power is transmitted from a motor. Double-drive systems are used when heavy duty pumping is required. With the single-drive gear, the second gear will idle.

Fluid is trapped in the cavity between the gear teeth and casing and transported around the outside of the gears to the delivery, where it is discharged. Removing a volume of fluid from the suction causes a drop in pressure that will induce fluid flow into the gear teeth to allow the process to continue.

These pumps are easily reversible since the direction of rotation determines the discharge side of the pump. The low pressure at the suction side makes them self-priming.

The form of gear used varies from simple spur gears to helical gears. Spur gears are frequently used in small pumps. When flow above 50 m³/h is required, the noise and thrust become very high and helical gears are used. At 1800 rpm, pressures from 3.0 to 3.5 MPa can be generated.

5.6.1.1.2 Internal Gear Pumps

Internal gear pumps have gentler fluid handling characteristics. In this pump the internally geared wheel meshes with an externally geared wheel. Fluid is trapped by the teeth and transported from the suction to the delivery side (see Figure 5.8).

The advantages of the gear pumps are as follows:

1. The pumps are self-priming.
2. The discharge is uniform with little pressure pulsation.
3. Pump rotation is reversible.
4. They are self-priming and can handle gaseous fluids.

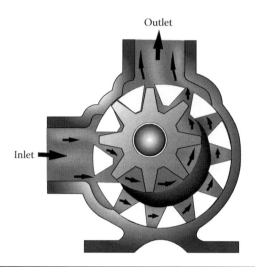

Figure 5.8 Internal gear pump.

5. The small clearances cause minimal flow variations with viscosity variations and pressure fluctuations.

The disadvantages of this type of pump are as follows:

1. Since the gear teeth have almost no clearances, only fluids with no solids can be pumped.
2. They can never be run dry.
3. Operation requires very close tolerances; alignment is therefore critical.
4. A pressure relief valve is required upstream from the pump.

Gear pumps are frequently used to pump light oil, boiler feed water, and products with lubricating properties, and for oil circulation.

5.6.1.1.3 Lobe Pumps

Lobe pumps operate in a similar way except that the lobe-shaped rotor tips intermesh. There is a very small clearance between the rotors and between the rotors and the casing.

The rotors are driven independently through timing gears. The trilobe form is normally used, while the bilobe form is reserved for difficult to handle fluids (Figure 5.9). These pumps are normally made in stainless steel for the food and pharmaceutical markets.

The pumps can pump at flow rates up to 800 m³/h with pressures of up to 2 MPa. The lobe rotors are manufactured from a variety of materials dictated by the application. The pumps are easy to clean and can handle shear, sensitive materials.

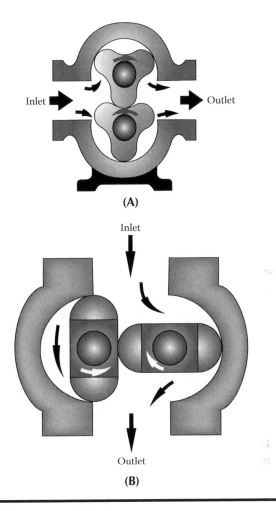

Figure 5.9 (A) Trilobe and (B) bilobe rotary pumps.

Advantages of this type of pump are as follows:

1. The design is hygienically approved.
2. If the rotors have a small clearance, the pumps can run dry for short periods of time.
3. They are self-priming when the lift is limited.
4. They can handle solids in suspension.
5. There is negligible velocity change as the product passes through the pump.
6. The flow is nonpulsating and reversible.

Disadvantages of this type of pump are as follows:

1. When the rotors have clearance and the pump is operated at low speed, the prime can be lost if the lift is high.
2. Fixed clearance can cause abrasive wear.
3. High pressure necessitates the use of special bearings to prevent shaft deflection, which could cause rotor damage.
4. A pressure relieve valve must be installed upstream.

5.6.1.1.4 Peristaltic Pump

This is the simplest form of positive displacement pump with very wide applicability (Figure 5.10). A flexible tube is laid in the curved track of the pump. The central rotor can have two or more rollers set at the periphery. As the rotor turns, the advancing roller progressively squeezes the tube driving the fluid before it. The portion of the tube behind the roller returns to its normal shape, thus creating low pressure to draw more fluid into the tube, which will be driven forward by the following roller.

These pumps are used in small-scale operations where capacities up to 10–12 m^3/h and relatively low pressures of up to 200 kPa are required. The durability of the tube is a problem and special reinforced elastomers offering high chemical resistance at elevated temperatures are used.

There is an industrial design that can achieve heads of 1 to 1.5 MPa and flow rates up to 75 m^3/h. The pumping action is gentle and large pieces of solids can be handled in the pumps.

Advantages of this type of pump are as follows:

1. The pump is self-priming.
2. It can be run dry.
3. It is reversible.

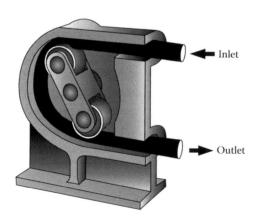

Inlet

Outlet

Figure 5.10   Peristaltic pump.

4. It will pump gaseous fluids.
5. It can pump abrasive materials.
6. It has only one contact part.
7. There are no glands or mechanical seals.
8. The tube can be sterilized for food and pharmaceutical operations.

Disadvantages of this type of pump are as follows:

1. Tube failure causes a big mess.
2. It cannot run against a closed valve.
3. The pressure is limited to the tube strength.
4. Pulsation can be a problem.
5. There is a relatively high maintenance cost of replacement tubes.

5.6.1.1.5 Progressive Cavity Pump

This type of pump has a screw revolving within a close-fitting barrel made from elastic material. The meshing of the screw and barrel cause the two parts to open and close a cavity that is passed from the suction to delivery side of the pump (Figure 5.11). As with the gear pump, the trapping and moving of a volume of fluid creates a low pressure on the suction side that induces fluid to flow into the pump.

Progressive cavity pumps are made from the materials that will best suit the application. Cast iron and stainless steel are frequently used. The resilient stator is manufactured from materials such as nitrile rubber. A capacity of up to 250 m^3/h with pressures up to 6 MPa is available. These pumps can handle liquids at temperatures up to 180°C.

The advantages of progressive cavity pumps are as follows:

1. They are self-priming.
2. Discharge is uniform with little pulsation.

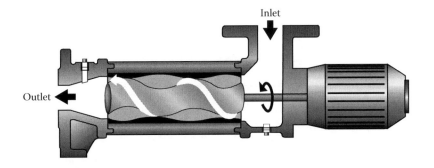

Figure 5.11 Helical screw pump.

3. The pumps are reversible.
4. They can pump gaseous fluids.
5. The pumps handle shear sensitive fluids with little damage.
6. They can handle solids in suspension.

The disadvantages of progressive cavity pumps are as follows:

1. They cannot be run dry without damage.
2. Since it is a positive displacement pump, it needs a pressure relief device.

5.7 Pipelines in the Industry

Piping should receive careful consideration in all industries. For this particular section, more emphasis will be placed on food industries since their piping requirements add the dimension of sanitation.

In the food industry, the final product must not only have all the required characteristics of color, viscosity, textural appearance, and many others, but it must also be clean and free of microbial contamination. Incorrect pumping is a problem that frequently rears its ugly head when a change is made in raw materials or when a new line is brought into production. Most problems with new lines center around pipelines and pumps.

5.7.1 Piping Materials

Industry is regulated by many codes of practice. For piping, the American National Standards Institute (ANSI) has codes covering piping in all steam-generating stations (ANSI B31.1), fuel gas piping (ANSI B31.2), refrigeration piping (ANSI B31.5), and many others. In certain services, the specific requirements for piping have been promulgated as part of the Occupational Safety and Health Act (OSHA) regulations. These codes will give minimum guidelines for safe piping installations and are not supposed to override common sense.

Ferrous metal piping is the most commonly used pipe and best described in national standards. They include wrought carbon and alloy steels such as stainless steel. Food plants use three types of pipe: standard pipe for water, gas, and ammonia; sanitary pipe; and copper tubing.

Standard pipe is usually made of wrought iron or steel. Ammonia pipe is nearly always made of steel. Low-pressure refrigeration piping is usually of copper tubing, with special soldered fittings. Gas pipes are frequently referred to as black pipes.

Pipe size is the approximate inside diameter (ID), with small sizes considerably oversize. Thirty centimeter (12 in.) or larger is listed outside diameter (OD). Pipe varies as to weight, with standard used for gas or water and extra strong for

Table 5.2   Nominal Pipe Sizes and Dimensions

Nominal Pipe Size		Outside Diameter (OD)		Inside Diameter (ID)		Nominal Wall Thickness	
in.	*mm*	*in.*	*mm*	*in.*	*mm*	*in.*	*mm*
0.50	12.70	0.84	21.34	0.62	15.80	0.109	2.77
0.75	19.05	1.05	26.67	0.82	20.93	0.113	2.87
1.00	25.40	1.32	33.40	1.05	26.64	0.133	3.38
1.50	38.10	1.90	48.26	1.61	40.89	0.145	3.68
2.00	50.80	2.38	60.33	2.07	52.50	0.154	3.91
2.50	63.50	2.88	73.03	2.47	62.74	0.203	5.16
3.00	76.20	3.50	88.90	3.07	77.93	0.216	5.49
4.00	101.60	4.50	114.30	3.17	80.52	4.026	102.26
6.00	152.40	6.63	168.28	6.07	154.05	0.28	7.11
8.00	203.20	8.63	219.08	7.98	202.72	0.322	8.18
10.00	254.00	10.75	273.05	10.02	254.51	0.365	9.27
12.00	304.80	12.75	323.85	11.04	280.37	0.406	10.31

high-pressure steam or ammonia and double extra strong for use with extremely high pressures (see Table 5.2).

For small diameter pipes, 50 mm or smaller, joints are frequently made with threaded fittings. The joint should be made so that there is very little or no change in pipe diameter. The joints should also be free of crevices that will accelerate corrosion and cause sanitary problems. For normal operations, beaded fittings will give increased strength. Fittings for steam and ammonia are normally made of cast iron and of malleable iron for water. Ammonia pipe and fittings are special, and ordinary standard pipe should not be used.

5.7.1.1 Pipe Assembly

Special care should be taken in pipe assembly to prevent leaks. The cutting of the thread in the pipe has weakened it and overtightening can cause failure of the pipe. The pipe threads do not match very well and a filler material called "pipe dope" should be used to prevent spiral leakage.

Brass and copper tubing connections are very widely used, since they are non-corrosive, simple to install, and can be easily bent, thus eliminating the use of many fittings as well as reducing the friction in the pipeline. Water, air, and low-pressure refrigeration lines are most adapted to the use of this type of piping.

5.7.2 Stainless Steel Pipeline

It is important that pipelines in the dairy, brewery, and beverage industries are sanitary. The stainless steel pipelines used in the sanitary pipe systems are thin walled and seam welded. The pipe is smooth enough to be easily cleaned. Stainless steel tubing is normally grade 304, but this grade is not resistant to corrosive solutions. When solutions containing sulfur dioxide are pumped, a special grade is required. Stainless steel tube is available in imperial and metric sizes (see Table 5.3).

The modern practice is to use the cleaning-in-place (CIP) system of sanitizing. Therefore, the design of the sanitary pipe system takes on added importance. It must be carefully worked out and engineered for both production and sanitizing. Proper drainage and circulation are very important.

The range of tubing parts includes all nominal standards as well as a range of pipe supports. All the tubing parts are manufactured out of high-quality pipes to accommodate customer demands for tolerances and overall quality.

5.7.3 Welding of Sanitary Pipe

A skilled operator should weld stainless steel, since it is difficult to obtain a smooth, nonporous welded surface inside the pipe. The parts should fit together and there should be no irregular surfaces or angles. When stainless steel tube is welded, there should not be any roughness or slag left on the insides of the tubes.

When pipelines are constructed, the following should be kept in mind:

1. Pipes must be free from crevices and pitting.
2. All pipes must drain.
3. Pipes must be properly supported.
4. Pipes must be correctly sized with cleaning in mind.
5. When pipelines need to be disconnected regularly, this ability should be incorporated into the design.

5.8 Pipeline Construction

Metals expand and contract when temperatures change. In a pipeline where temperature fluctuations occur, the change in total length of the pipeline must be considered during installation. Flexible rubber joints are frequently used on the suction and discharge flanges of centrifugal pumps. These joints perform one or more of the following functions:

Table 5.3 Dimensions of Stainless Steel Tubing

Imperial (inch)				Metric (millimeter)			
Nominal Tube Size	Outside Diameter	Inside Diameter	Wall Thickness	Nominal Tube Size	Outside Diameter	Inside Diameter	Wall Thickness
1.00	1.00	0.90	0.050	NW 10	12.00	10.00	1.0
1.50	1.50	1.40	0.050	NW 15	18.00	16.00	1.0
2.00	2.00	1.90	0.050	NW 20	22.00	20.00	1.0
2.00	2.00	1.87	0.065	NW 25	28.00	26.00	1.0
2.50	2.50	2.37	0.065	NW 32	34.00	32.00	1.0
3.00	3.00	2.87	0.065	NW 40	40.00	38.00	1.0
4.00	4.00	3.83	0.065	NW 50	52.00	50.00	1.0
				NW 50	52.00	48.80	1.6
				NW 65	70.00	66.00	2.0
				NW 80	85.00	81.00	2.0
				NW 90	93.00	89.00	2.0
				NW 100	104.00	100.00	2.0

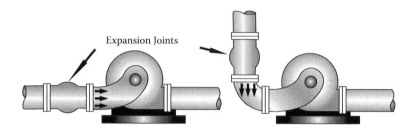

Figure 5.12 Expansion joints incorporated in pipeline close to centrifugal pumps.

1. Accommodate thermal expansion or other pipeline movement
2. Reduce vibration transfer from the pipeline to the pump and vice versa
3. Take care of misalignment between pump and piping
4. Facilitate maintenance on the pipeline without disturbing the pump
5. When the pipeline and pump have different flanges, act as a flange adaptor
6. When a pump is mounted on antivibration mountings, allow freedom of movement
7. Reduce noise transmission

When a pipeline is under pressure, the hydraulic thrust acts on the pipe ends and at each change of direction. The strength of the pipe prevents it from stretching. The rubber expansion joint removes the rigidity and the stiffness of the joint holds the pipe together (Figure 5.12). In pipelines where rubber expansion joints cannot be used, horseshoe bends are incorporated to deal with expansion and contraction.

Devices such as tie rods, anchors, and supports must control the thrust on the pump. Remember the pump should never be used as a support for the pipeline.

Problems

1. Calculate the power (kilowatts) required to pump 100 L/min of water from a container to an elevation 12 m above the free surface of the water in the container. The pump is connected to the container by 1 in. standard iron pipe. The pipeline is constructed as follows: Distance between container and pump is 125 cm, vertical pipe 12 m, one 1 in. elbow, and 5 m of horizontal pipe. The pump efficiency is 70 percent.

$$\text{Water density} = 1000 \text{ kg/m}^3$$

$$1 \text{ in. pipe} = 26.64 \text{ mmID} = Dm$$

$$L = 1.25 + 12 + 0.84 + 5 = 19.09 \text{ m}$$

$$Q = \frac{100 \text{ L}}{\text{min}} = \frac{1.667 \text{ L}}{\text{s}} = Q \times 10^{-3} \text{ m}^3/\text{s}$$

$$A = \frac{\pi}{4}(Dm)^2 = A \times 10^{-4} \text{ m}^2$$

$$V = \frac{Q}{A} = \frac{Q \times 10^{-3} \text{ m}^3/\text{s}}{A \times 10^{-4} \text{ m}^2} = V \text{ m/s}$$

$$\text{Re} = \frac{DV\rho}{\mu} = \frac{Dm \times V \text{ m/s} \times 1000 \text{ kg/m}^3}{9.8 \times 10^{-4} \text{ Pa} \cdot \text{s}} = \text{Re}$$

The flow is turbulent since the Reynolds number is close to 100,000. Do the calculations. Please remember that Pa · s = kg/m · s.

$$\text{Friction factor } f = 0.079 \text{Re}^{-0.25} = \frac{0.079}{\sqrt[4]{\text{Re}}} = FF \times 10^{-3}$$

$$F = 2f\frac{LV^2}{Dg} = 2 \times FF \times 10^{-3}\left(\frac{19.09 \text{ m} \times (V \text{ m/s})^2}{Dm \times 9.81 \text{ m/s}^2}\right)$$

$$F = X \text{ m} = X \text{ Nm/N} = X \text{ J/N} = X \text{ W}$$

$$h = 12 + 6.110 = 18.11 \text{ m}$$

$$P_0 = WQ\rho g = 18.110 \text{ m} \times Q \times 10^{-3} \text{ m}^3/\text{s} \times 1000 \text{ kg/m}^3 \times 9.81 \text{ m/s}^2$$

$$P_0 = PP \text{ kgm}^2/\text{s}^3 = PPW = PPkW$$

$$e_m = 0.70 = \frac{PPW}{W}$$

$$0.70 \text{ W} = PPW$$

$$\text{Pump} = YkW$$

2. In a processing plant, water at 15°C is pumped at a rate of 80 L/min through a horizontal length of ¾ in. standard galvanized pipe. The pressure drop in this setup is found to be 600 kPa. To reduce the pressure drop, the plant engineer proposed that the pipe be replaced with 1 in. pipe. Calculate the pressure drop if the flow volume of 80 L/min is maintained. (Hint: since the length of

pipe is unknown but stays the same, it will cancel out in both equations. In fact, a number of things stay the same and can be canceled.)

3. Calculate the power rating of the pump necessary to pump fluid from a reservoir to a tank 15 m higher at a rate of 200 L/min. Assume the stainless steel pipe has an inside diameter of 26 mm, the viscosity is 1 cP, fluid density is 1.05 g/mL, and the fluid velocity in the reservoir is zero. Disregard friction; the pressure inside the reservoir and the tank is 1 atm and the pump efficiency is 65 percent.

Chapter 6

Thermodynamics

Thermodynamics is the study of the relationship between heat and work. The first law of thermodynamics is merely a quantitative expression of the principle of conservation of energy. This means that you cannot get more energy out of a system than the total energy that went into the system. Since all materials have energy associated with them, mass as well as energy must be taken into account. This law is of importance where transfer of mass or energy is involved.

The second law states that the transfer of energy by heat flow is directional and not always reversible. The flow of thermal energy is always from higher to lower energy levels.

6.1 Energy of a System

Energy or work is force multiplied by distance moved in the direction of the force. It is measured in joules (J). In any thermodynamic system, energy cannot be created or destroyed; it can only be converted from one form to another. The accounting of energy can only be done when a system is defined and delineated. The system must have boundaries that can be real, virtual, or partially real and virtual. The region outside the system will be called the surroundings. In a closed system, only energy is interchanged and mass stays constant. In an open system, both mass and energy will interchange.

A steady-state flow system is a special kind of open system. Mass enters and leaves the system at the same rate and although temperatures and pressures may vary in different parts of the system, they do not change with time. An example of this kind of system is a plate heat exchanger. This kind of system is the one most frequently encountered in a processing line.

Energy is never static; there is always some interchange. Even in a system where the mass and energy stay the same (steady-state condition) some energy will be absorbed from the surroundings and some will be emitted to the surroundings. The absorbed energy will equal the released energy.

6.2 Characteristics of a System

Any characteristic of a system that is reproducible is a *property* of such a system. The *intensive* properties of a system do not depend on mass and are therefore independent of the size of the system. Intensive properties include temperature, pressure, density, and viscosity. Water will boil at 100°C at standard pressure and is not affected by the size of the container. Volume, on the other hand, is an *extensive* property since it correlates directly with mass. Two kilograms of water has twice the volume of 1 kg of water at the same temperature and pressure. If we divide the volume by the mass, the *specific density* in cubic meters per kilogram will be the same, regardless of the mass. It is therefore possible to change the extensive properties of a system to intensive properties by dividing the magnitude of the property by its mass.

The *state* of a system is its condition at specific property values. When a system is allowed to pass through a variety of changes and is then brought to a condition in which its properties are the same as they were at some previous time, it is in the same state. When a *process* takes place, the system changes from one state to another.

6.2.1 Energy of a System

Thermodynamic energy can be divided into quantities of energy associated with the state of the system; they are properties and quantities of energy, which are recognized when the system is undergoing a process.

External energy can be considered as kinetic energy and gravitational potential energy that a system possesses because of its velocity or position. The value of this energy is relative to some datum, which can be chosen as a reference point. When an appropriate frame of reference is chosen, changes in external energy can be eliminated.

Internal energy is associated with the atomic and molecular structure of the substance. No frame of reference can be chosen such that it will eliminate the changes in internal energy. Internal energy could be in the form of nuclear, chemical, thermal, or molecular energy. Nuclear energy is associated with the disintegration energies of nuclear particles. Chemical energy is part of the bonding energies of atoms within molecules. Thermal energy is related to translation, rotation, vibration, and electron shifts of molecules and is directly related to temperature. Molecular energy is associated with attractive and repulsive forces among molecules.

A substance possesses a specific amount of internal energy irrespective of its state. When an amount of heat, Q, is allowed to flow into the system and W is the net amount of work done by the system, then the conservation of energy gives

$$Q = W + \Delta U$$

where ΔU is the change in internal energy $(U_2 - U_1)$. Rearrangement of the equation gives

$$\Delta U = Q - W$$

When Q is positive, heat is added to the system and if it is negative, it means that the system is giving off heat. In the same way, a negative value for W will mean that work is performed on the system rather than being done by the system. The change in internal energy of the system is equal to the difference between the heat transfer in the system and the work being done by the system.

If no heat enters or leaves a system during a process, then the process is assumed to be perfectly isolated from its environment and called adiabatic. In this case $Q = 0$. This is an ideal situation since some heat is always transferred in actual systems. Processes that are almost adiabatic include the rapid compression of air in a tire pump. The equation then becomes

$$\Delta U = -W$$

In an adiabatic process, the system does not exchange heat with the environment but undergoes a change in internal energy that is the negative value of the work done by the system.

When a process takes place and there is no change in temperature, the process is termed isothermal. These processes normally proceed so slowly that the rate of temperature change is very small.

$$Q = W$$

Very few processes will be adiabatic or isothermal, although treating them as one or the other, or one following the other, will give a reasonable approximation.

A process that takes place without a change in volume is called an isochoric process. In such a process, no displacement can take place, so the work done by the system is zero.

$$Q = \Delta U$$

In an isochoric process, the system does no work; therefore, all the energy that is added to the system will increase the internal energy of the system.

Table 6.1 Critical Temperatures for Some Substances

Compound	Molar Mass (kg/kmol)	Critical Temperature (°C)	Critical Temperature (K)
Air	28.95	−140.7	132.45
Oxygen	32.0	−118.57	154.58
Nitrogen	28.01	−146.95	126.2
Ammonia	17.03	132.5	405.65
Carbon dioxide	44.01	31.06	304.21
Water	18.02	373.98	647.13

Source: Adapted from a table in Perry, R. H. and D. W. Green. 1997. *Perry's Chemical Engineers' Handbook,* 7th ed. New York: McGraw–Hill.

6.3 Gases and Vapors

Substances that are normally in the form of a gas at ambient temperatures are called gases. The term vapor describes the gaseous phase of substances that are liquids or solids at ambient temperatures. For pure substances, another distinctive difference between gases and vapors is that gases normally contain only gas, while vapors could contain a mixture of gas and liquid. In some cases, the terms could also be synonymous.

When a gas is compressed, it will liquefy at a specific pressure only if the temperature is lower than a specific critical temperature for that gas. The more the temperature is reduced below the critical temperature, the easier it is to liquefy the gas. Above the critical temperature, the gas will not liquefy no matter how high the pressure becomes.

Substances with critical temperatures above 50°C are normally called a vapor, while those with critical temperatures below 50°C are called gases (see Table 6.1).

If we consider each molecule of a gas as a discrete particle, then the collision between these particles can be described by the laws of mechanics. The characteristics of the gas such as pressure and internal energy can thus be described in terms of molecular motion. Using a molecular count yields large numbers that are meaningless in a practical sense. One can weigh 1 kg of a substance to use in a formulation, but adding fifty zillion molecules can be a bit of a bother unless we convert the number of molecules to mass.

6.3.1 Ideal Gas Equation

The following basic assumptions are made to postulate the kinetic theory of gases:

- In a pure gas, all the molecules will have the same mass, which is so small that gravity does not affect it. It is also in a state of complete random motion.
- The gas molecules have very large distances between them, so the volume that the molecules occupy is negligibly small in comparison to the total volume wherein the gas exists.
- There are no intermolecular forces except when collisions occur.
- Collisions between molecules and between the molecules and the walls of the container are perfectly elastic.

The force of these collisions with the walls of the container can be calculated with Newton's laws of motion. When force is expressed as pressure per area we get

$$PV = \frac{1}{3} Nmv_{rms}^2$$

where P is the pressure, V is the volume of the container of gas, N is the number of gas molecules in the container, m is the mass of one gas molecule, and v_{rms} is the root mean square of the velocity of the molecules.

When the gas laws as postulated by Boyle and Charles were combined, the following relationship between pressure (P), volume (V), and temperature (T) emerged. This is called the ideal gas law:

$$\frac{P_1 V_1}{T_1} = \frac{P_2 V_2}{T_2}$$

The temperature in the equation is absolute temperature (Kelvin, K). Remember that 0°C is equal to 273.2 K. For calculation purposes, we normally use only the 273 and drop the 0.2. Absolute temperature can be calculated by adding 273 to the Celsius temperature.

6.3.2 Pressure Expressions

Pressure can be expressed as absolute pressure. This is the pressure that is exerted by the gas molecules colliding with a surface. It is thus proportional to the number of molecules and the energy of the molecules. The units for this pressure are kilopascals.

Pressure can also be expressed using atmospheric pressure as units. The atmosphere refers to a standard pressure of 101.325 kPa that is the normal atmospheric pressure at sea level. When we set a gauge to zero at sea level, pressure can be expressed as units above the pressure at sea level; this is called gauge pressure.

$$P_a = P_g + P_{atm}$$

(Absolute pressure = gauge pressure + atmospheric pressure)

When the gauge pressure is negative, it means the pressure is lower than atmospheric. We refer to that as "vacuum." Vacuum is the amount that the pressure is below atmospheric pressure and it is a positive value.

$$P_{vac} = P_{atm} - P_a$$

The lowest possible absolute pressure that is theoretically possible is zero. Since pressure occurs when molecules collide with the walls of a container, zero pressure means complete absence of molecules.[*] The largest theoretical vacuum is therefore equal to atmospheric pressure. At this point, the gauge pressure will be −101.325 kPa at standard temperature and pressure (STP).

Example 6.1

In food processing, cans are sealed with a small headspace. Consider what will happen when an empty can enters a retort:

$$T_1 = 20°C = 273 + 20 \text{ K}$$

$$T_2 = 140°C = 273 + 140 \text{ K}$$

$$P_1 = 101.3 \text{ kPa}$$

Assume no change in volume of the can:

$$\frac{P_1 V}{T_1} = \frac{P_2 V}{T_2}$$

$$P_2 = \frac{P_1 T_2}{T_1} = \frac{101.3 \text{ kPa} \times (140 + 273) \text{K}}{(20 + 273) \text{K}}$$

$$P_2 = 142.8 \text{ kPa}$$

This will explain why empty cans and even half-filled cans frequently explode in a retort.

---

[*] At 0 K, −273°C, all molecular movement stops, thus causing a loss of pressure.

Example 6.2

A can full of air is sealed at atmospheric pressure and 80°C. What will the gauge pressure be after the can is cooled down to 20°C?

$$\frac{P_1 V}{T_1} = \frac{P_2 V}{T_2}$$

$$P_2 = \frac{P_1 T_2}{T_1} = \frac{101.3 \text{ kPa} \times (20 + 273)\text{K}}{(80 + 273)\text{K}}$$

$$P_2 = 84.1 \text{ kPa}_a$$

$$P_g = 84.1 - 101.3 \text{ kPa}$$

$$P_g = -17.2 \text{ kPa}$$

In this case, the vacuum will be 17.2 kPa. When the can starts denting, the volume will change and the pressure will normalize. If the can was dented to the extent that the pressure was steady, what would the volume be if the original volume were 800 cm³?

$$\frac{P_1 V}{T_1} = \frac{P_2 V}{T_2}$$

$$V_2 = \frac{V_1 T_2}{T_1} = \frac{800 \text{ cm}^3 \times (20 + 273)\text{K}}{(80 + 273)\text{K}}$$

$$V_2 = 664 \text{ cm}^3$$

6.3.2.1 The Perfect Gas Law

The ideal gas law applies to a specific quantity of gas. When we want to work with any other quantity of gas we have to include a measure for the amount of the gas in the equation. The relationship, called the perfect gas law or the ideal gas law, is written to suit any dilute gas:

$$\frac{PV}{T} = Nk_B = \frac{1}{3} Nmv_{rms}^2$$

$$3Nk_B = Nmv_{rms}^2$$

N is the number of molecules and k_B is the Boltzmann's constant that equals 1.38×10^{-23} J/K. Remember that the temperature is given in kelvin.

The number of molecules in a sample can be a rather cumbersome number to work with, so the ideal gas equation was formulated to operate with molar quantities of gas and Boltzmann's constant was replaced with the universal gas constant R = 8.31 J/(mol·K).

$$\frac{PV}{V} = nR$$

$$PV = nRT$$

Remember that the temperature is on the kelvin scale. The mole (abbreviated as mol) is defined as a quantity that contains Avogadro's number of molecules (6.02 × 10²³ molecules/mole).

$$n = \frac{N}{N_A} \quad \text{and} \quad n = \frac{m}{M}$$

The number of moles (n) is equal to the total number of molecules (N) divided by Avogadro's number (N_A). The number of molecules is obviously dependent upon the mass of the substance, so n can also be defined as the mass divided by relative molecular mass.

Example 6.3

What is the speed of a nitrogen molecule (N_2) in a gas-flushed package at room temperature? Molar mass of nitrogen (N) is 14 g/mol; molar mass of N_2 is 28 g/mol.* Assume that all other gasses were replaced by nitrogen.

First you must calculate the mass of one molecule of nitrogen:

$$m_N = \frac{\left(\frac{28\ g}{mol}\right)\left(\frac{kg}{1000\ g}\right)}{6.02 \times 10^{23}\ molecules/mol} = 4.65 \times 10^{-26}\ kg/molecule$$

Room temperature is 20°C = 293 K and Boltzmann's constant = 1.38 × 10⁻²³ J/K:

$$v_{rms} = \sqrt{\frac{3k_B T}{m}}$$

$$v_{rms} = \sqrt{\frac{3 \times 1.38 \times 10^{-23} \times 293}{4.65 \times 10^{-26}}} = \sqrt{260864.5} = 510.7\ m/s$$

At a temperature of 40°C the velocity would have been 527.9 m/s. At this higher velocity, the molecules will move from one wall of the container to another wall in a shorter time. This causes more collisions with the walls of the container and results in a higher pressure.

---

* Remember the seven diatomic elemental gases: hydrogen, nitrogen, oxygen, fluorine, chlorine, bromine, and iodine.

By using this equation, the volume of 1 mol of a gas at STP (0°C, 273 K; 1 atm) can be calculated as 22.4 L. The volume of 1 kmol will be 22.4 m³.

6.3.3 *Mixtures of Gases*

The perfect gas laws work for most of the monatomic and diatomic gases at *low* pressure and *high* temperatures, well above the critical temperature of the gas. The gas laws can be used for most of the general conditions within the manufacturing environment.

The perfect gas laws can also be used for mixtures of gases. If we define n_t as the sum of moles of all gases, M as the concentration in moles of each gas, m_t as the total mass of all gases, and M_{ave} as the average molecular mass, we get

$$m_t = m_1 + m_2 + \cdots = M_1 n_1 + M_2 n_2 + M_3 n_3 \cdots$$

$$n_t = n_1 + n_2 + \cdots$$

$$M_{ave} = \frac{m_t}{n_t} = \left(\frac{M_1 n_1}{n_t} + \frac{M_2 n_2}{n_t} + \frac{M_3 n_3}{n_t} + \cdots \right)$$

For practical purposes, we assume that air contains 0.21 mol fraction oxygen and 0.79 mol fraction nitrogen. The average molar mass of air is 29.0 g/mol or 29.0 kg/kmol.

The gases and vapors that are present in a mixture of gases will exert pressure independently of each other. Dalton's law states that the total pressure of a mixture of gases and vapors is equal to the sum of the partial pressures of each component of the mixture:

$$P_t = P_1 + P_2 + P_3 + \cdots$$

The partial pressure of a specific gas in a mixture is the pressure that the gas would exert if it alone occupied the space. In air, the partial pressure consists of the partial pressure of air plus the partial pressure of water vapor. The total partial pressure of air is the sum of the partial pressures of oxygen, nitrogen, carbon dioxide, hydrogen, and the other gases.

In much the same way, one can work with partial volumes. Amalgat's law of partial volumes states that the total volume of a mixture is the sum of the volumes of the components of the mixture:

$$V_t = V_1 + V_2 + V_3 + \cdots$$

The ideal gas equation can be used for each component gas and vapor:

$$P_1 V = n_1 RT \quad \text{or} \quad PV_1 = n_1 RT$$

Composition of gas mixtures is given as volume percentages of each constituent gas.

Example 6.4

Calculate the density (kilograms per cubic meter) change in air when the temperature changes from 20°C to 100°C. The pressure is constant at 1 atm. Density is mass/volume.

$$PV = nRT = \left(\frac{m}{M}\right) RT$$

$$V = \frac{mRT}{MP}$$

$$\rho_{293\,K} = \frac{m}{V} = \frac{m}{1} \times \frac{MP}{mRT} = \frac{MP}{RT} = \frac{29.0 \text{ kg/kmol} \times 101.3 \text{ kPa}}{8.31 \text{ Pa} \cdot \text{m}^3/(\text{kmol} \cdot \text{K}) \times 293 \text{ K}} = 1.21 \text{ kg/m}^3$$

$$\rho_{373\,K} = 0.95 \text{ kg/m}^3$$

The density will change from 1.21 to 0.95 kg/m³.

Does this make sense? If there is 1.21 kg air in 1 m³ at 293 K, then there will be the same mass of air at 373 K, but if the pressure stays the same the volume will change. From the ideal gas equation, the new volume would be

$$V_2 = \frac{V_1 T_2}{T_1} = \frac{1 \text{ m}^3 \ 373 \text{ K}}{293 \text{ K}} = 1.273 \text{ m}^3$$

$$\rho = \frac{m}{v} = \frac{1.21 \text{ kg}}{1.273 \text{ m}^3} = 0.95 \text{ kg/m}^3$$

It is always good to check your answers to ensure that the answer is at least in a possible range.

Example 6.5

A 1 kg block of solid carbon dioxide (molar mass: 44 kg/kmol) is placed in a container with a volume of 0.5 m³. What would the pressure be after the container and contents have been equilibrated to 22°C? When the dry ice was placed inside the container, the pressure was 101.3 kPa. Since the volume of the block of carbon dioxide is about 1 L and the container has a volume of 500 L, the change in volume of the air can be disregarded.

To use the ideal gas equation, first calculate the molecular concentration of the gas.

$$n_{CO_2} = \frac{1 \text{ kg}}{44 \text{ kg/kmol}} = 0.0227 \text{ kmol}$$

Now calculate the pressure.

$$P_{CO_2} = \frac{nRT}{V} = \frac{0.0227 \text{ kmol} \times 8.32 \text{ kJ/kmol} \cdot \text{K} \times 295 \text{ K}}{0.5 \text{ m}^3}$$

$$P_{CO_2} = 111.43 \text{ kPa}$$

From Dalton's law we get

$$P_t = P_{air} + P_{CO_2} = 101.3 + 111.43 \text{ kPa} = 212.73 \text{ kPa}$$

Gas occupies a much larger space than the solid or the liquid from which it vaporized. It is natural to increase the pressure.

Example 6.6

A half-empty 1 L bottle of water is left at 20°C. Calculate the quantity of air in the 0.5 L (500 cm³) volume of gas and vapor.

The metric steam table in Appendix 3 gives the vapor pressure for water at 20°C as 2.3366 kPa.

Since the bottle is half empty, assume that it is at atmospheric pressure, 101.3 kPa:

$$P_{air} = P_t - P_{vapor} = 101.3 - 2.34 = 98.96 \text{ kPa}$$

Using the ideal gas equation and solving for n,

$$n_{air} = \frac{P_{air}V}{RT} = \frac{98.96 \text{ kPa} \times 0.5 \times 10^{-3} \text{ m}^3}{8.32 \text{ kJ/kmol} \cdot \text{K} \times 293 \text{ K}}$$

$$n_{air} = 2.03 \times 10^{-5} \text{ kmol}$$

Example 6.7

The partial pressure of water in saturated air at 20°C and 1 atm is 2.3366 kPa. If the air is compressed to 10 atm at a temperature of 30°C, what would the partial pressure of the water in the compressed air be?

The partial pressure for

$$n = \frac{P_1 V_1}{RT_1} = \frac{P_2 V_2}{RT_2}$$

Since n and R do not change, the equation can be reduced to

$$\frac{V_1}{V_2} = \frac{P_2 T_1}{P_1 T_2} = \frac{10 \text{ atm} \times 293 \text{ K}}{1 \text{ atm} \times 303 \text{ K}} = 9.67$$

When the equation is rewritten for vapor pressure, the following equation is derived:

$$\frac{V_{1vap}}{V_{2vap}} = \frac{P_{2vap} T_1}{P_{1vap} T_2} = \frac{P_{2vap} \times 293 \text{ K}}{2.3366 \text{ kPa} \times 303 \text{ K}} = 9.67$$

$$P_{2vap} = \frac{9.67 \times 2.3366 \text{ kPa} \times 303 \text{ K}}{293 \text{ K}} = 23.37 \text{ kPa}$$

The assumption was made that no vapor would condense. Looking at the steam table in Appendix 3, the vapor pressure at 30°C is 4.246 kPa. This means that some vapor would have condensed, and the partial pressure of the vapor is 4.246 kPa.

Problems

1. A cylinder of compressed air had a gauge pressure of 710 kPa at the start of an operation. After the operation the gauge pressure was 35 kPa. If the cylinder has a volume of 0.05 m³, what volume of air at atmospheric pressure was delivered? The operation was slow enough so that the temperature can be assumed constant.

2. Calculate the density change in air that is cooled from 90°C to 20°C at constant pressure of 101.3 kPa. Assume that the air was dry enough so that no condensation occurred.

3. Calculate the change in saturated air density when the air is cooled from 100°C to 20°C at a constant pressure of 101.3 kPa. Condensation will occur.

4. In an air cooling system, 200 m³ saturated air is moved over cooling coils operating at 0°C. The temperature of the air leaving the cooling system is 5°C. Assume that some condensation occurs on the coils, meaning that the vapor pressure of the air is equilibrated to 5°C. This air is mixed with 1000 m³ saturated air at 25°C. Calculate the final temperature of the air and the final moisture content.

5. A can of peanuts has a total volume of 884 mL and contains 435 mL peanuts. Calculate the moles of oxygen inside the can.

6. A bin drier containing 2000 kg of material with a moisture content of 50 percent wet mass basis has air leaving the drier at 60°C. The incoming air is normal saturated air that is heated to 85°C. How much air must be passed through the system to reduce the moisture content to 8 percent wet mass basis? Assume free movement of moisture out of the product.

Chapter 7

Electrical Systems

Industry relies on electricity as the energy source to control almost all the systems. In some cases, automated mechanical control or even manual control are still used but they are mainly in older plants or in specific situations where regulations still enforce the use of such controls. Most control rooms are sophisticated, with many video screens and banks of buttons and knobs. For the uninitiated it is definitely a hands-off situation.

Since electricity is essential for light, heat, energy, communications, and control, it is important that one has some knowledge of electrical basics. The information contained in this chapter will provide some helpful background to the nonengineer and nonelectrician working in industry. It should also provide basic information to make judgments regarding the use, performance, and purchasing of electrically operated equipment for the plant.

7.1 Electricity Generation

Electrical power starts at the generators in a power plant. The energy to drive the generator can be a water wheel in a hydroelectric system, a diesel engine for smaller operations, or steam. With steam generators, water is heated by burning coal, oil, or natural gas, or by nuclear reactions.

Commercial generators generate three-phase alternating current (AC) electricity. For household use, we use just single-phase 120 V AC service. The electricity will oscillate between +170 V and −170 V and the average voltage will be 120 V. The form of the peaks will be a sine wave and the rate of oscillation will be 60 cycles per second.

Power plants produce three different phases of power simultaneously. The three phases are offset 120° from each other. The three phases plus the neutral need to be carried by four wires. At any stage in the cycle, one of the phases will reach a peak. When using single-phase power, there are 120 moments per second when the sine wave crosses 0 V. At the stage when the voltage drops to zero, the power also drops to zero. When a single phase is used to drive a motor, the torque to the shaft of the motor will pulse from full power to zero 120 times per second. This will not be noticeable in small motors where torque is not very important. In industry, when torque is very important, three-phase motors are used.

The principal advantage of the AC power is that the voltage can be very easily and efficiently stepped up or down by means of a transformer to give the desired high voltage for efficient long-distance transfers of energy. At the user end, it can be stepped down to a safe usable voltage. The electricity is transported at voltages in the order of 155,000 to 765,000 V to reduce transfer loss. The electricity is then transformed at the user end to 120 or 220 V.

7.2 Definitions

Ampere is the term used for current. When an ammeter reads 1 ampere (A) then 1 coulomb (C) of charge is flowing every second:

$$1 \text{ A} = 1 \text{ C/s}$$

Volt is the term for potential difference between two points. It is the work done in joules to move 1 C of charge between the plates. The SI unit is joule/coulomb (J/C) or volt (V), where

$$1 \text{ V} = 1 \text{ J/C}$$

Potential difference is referred to as voltage and we also use V to designate it. In most plants, the service at the distribution end of the line is 110 or 220 V, although some large items, such as refrigerator compressors, may operate at 440 or even 2200 V in order to gain efficiency. Special precautions must be taken with any voltage over 220 and only limited, trained personnel should have access to the equipment.

Resistance: an object has a resistance of 1 ohm if a potential difference of 1 V across it will result in a current of 1 A to flow:

$$R = \frac{V}{I} \text{ohm} \quad \text{and} \quad V = IR$$

Watt is potential difference times the current. This is just an approximation since there is the matter of power factor, which should be considered.

$$W = V \times A = (V \times I) \text{Joule/second}$$

Substituting the formula for volts, we get

$$W = (I^2 \times R) \text{J/s}$$

The energy used in t seconds can be given by

$$W = (I^2 \times R \times t) \text{J/s}$$

Electrical meters measure energy consumption in kilowatt-hour (kWh). It is the energy supplied to a 1 kW appliance when it is connected to the supply for 1 h. A 1 kW appliance gets 1000 J/s of energy for 1 h—that is, 3600 s:

$$1 \text{ kWh} = 1000 \frac{J}{s} \times 3600 \text{ s} = 3.6 \times 10^6 \text{ Joule}$$

It is important that one considers the power factor of equipment. A power factor close to one denotes good efficiency of the electrical system and means that less heat is lost in the system. Heaters and resistance heaters normally have a power factor of one while partially loaded electric motors cause a decrease in power factor.

Insulators are materials that do not conduct electricity well. The materials include rubber, plastics, glass, and ceramics. Insulated wires and conductors are necessary in motors and controls and to insulate the electrical system to prevent dangerous shocks and loss of energy.

Conductors are materials such as copper, aluminum, and iron that conduct electrical energy with little loss. Copper is a very efficient and frequently used conducting material. It is expensive and is replaced by aluminum in special cases.

Phase is important to the user of electric motors. It applies to alternating current circuits and is mainly involved in motors.

Cycles refer to the reverse of the current in AC circuits. In the United States, the current is reversed sixty cycles or 120 times per second. When we connect a 60-cycle motor to a 50-cycle circuit, it will run only five-sixth of its rated speed. For example, a 120-rpm 60-cycle motor placed on a 50-cycle circuit will run only 100 rpm.

7.3 Energy Conversion

People frequently think of electricity as a clean source of energy with few pollution consequences. Consider the conversion of energy from coal to electricity. If we keep a 100 W light bulb on for a year, how much electricity will it consume and how much coal was burned to produce the electricity?

Energy used is

$$100 \text{ Watt} \times \frac{1 \text{ kW}}{1000 \text{ W}} \times \frac{24 \text{ h}}{\text{day}} \times 365 \text{ days} = 876 \text{ kWh}$$

Bituminous coal has an energy value of 30.2 MJ/kg. To calculate the energy per kilogram,

$$1 \text{ kg} = 30.2 \text{ MJ} \times \frac{1 \times 10^6 \text{ J}}{\text{MJ}} \times \frac{1 \text{ kWh}}{3.6 \times 10^6 \text{ J}} = 8.39 \text{ kWh}$$

Coal-fired power generators are 40 percent efficient; this means that only 40 percent of the available energy is converted to electricity. To keep the light bulb on for 1 year, we need to burn

$$\text{Coal (kg)} = \frac{876 \text{ kWh}}{8.39 \text{ kWh/kg}} \times \frac{1}{0.4} = 261 \text{ kg}$$

It is a sobering thought that we need to burn more than a quarter ton of coal to produce the energy to light one bulb for a year. The burning of the coal will result in waste products such as SO_2, CO_2, CO, and ash. There are a lot of nasty by-products associated with coal-generated electricity. The footprint actually starts with the mining of the coal.

As a comparison, consider what happens in a nuclear power plant. Uranium is mined, milled to concentrate and purify it, and then converted into metal uranium, uranium dioxide, or uranium hexafluoride. The uranium is then fabricated into rods or elements and these are loaded into the nuclear reactor of the power plant. In the reactor, nuclear fission occurs that produces thermal energy, just like coal, to produce steam to drive the turbines that generate electricity.

Once the fuel is spent, new fabricated rods or elements are loaded and the used fuel rods are the big problem. The spent material has to be kept cool while the major part of its decay takes place. The material is then reprocessed to recover any remaining uranium and the plutonium that was produced in the reactor. The material then has to be stored in remote, very secure, specialized storage facilities for thousands of years. Plutonium-239 has a half-life of 24,100 years. It can also sustain nuclear chain reactions and is therefore a prime component in nuclear weapons.

I sometimes wonder which method of power generation has the greatest environmental impact.

7.3.1 Electrical Energy

Electrical energy is sold as the amount of kilowatt-hours used. An additional amount is added to the bill and this is determined by factoring the size and uniformity of the load as well as the time of day the energy is used. A kilowatt-hour of electrical energy is equivalent to 1000 W flowing for 1 h, 500 W for 2 h, or 2000 W for 30 min.

Most plants purchase some or all of their electrical energy from a power company. In plants that generate excess steam part of the day, the excess steam will be used to generate electricity. Other plants will use diesel electric generators to augment their available electricity to reduce peak loads during start-up time. Peak load electricity in the morning is normally more expensive than at other times of the day. In-plant generators are also used as an emergency energy source in areas where public power is not very reliable or where winters frequently cause power grid problems.

7.4 Electric Motors

The size of the electrical motors used in industry is awesome and it follows that a very large amount of electrical energy is required to run them.

All electrical motors operate under the same principle. When current passes through a wire, the result is a magnetic field around the wire. The force on the wire is produced perpendicular to the current and the magnetic field. The magnetic field will radiate inward toward the rotor axle or outward toward stator. Conductors that are parallel to the axis will cause a force tangential to the rotor circumference. The situation will cause the rotor to turn.

7.4.1 Single-Phase Motors

7.4.1.1 Single-Phase Universal Motor

Single-phase motors are normally used in applications where less than 1 kW motors are required. These motors have good starting torque. Speed control on these motors is not very good since it is regulated with resistance. The direction of rotation of the field coil connections must be switched.

7.4.1.2 Single-Phase Split-Phase Motor

Single-phase split-phase motors are efficient but have low starting torques. They are used for driving fans and other equipment where no heavy loads are started and where less than 1 kW is required. The design incorporates a centrifugal switch to cut out a starting coil once the motor is up to speed. The rotor is usually of the squirrel-cage type. To change the direction of rotation, one needs to switch the starting coil terminal wires.

7.4.1.3 Single-Phase Capacitor Motor

This motor is used for fractional kilowatt services and in larger sizes only when three-phase current is not available. Appearance and construction are similar to split-phase motors. It has a condenser and resistance combination housed in a container attached to the motor, giving it some very desirable features. It is more expensive than the split-phase motor.

It has ideal starting and running characteristics. The capacitor motor will accelerate any load that the motor will carry because the start-up torque is as good as or better than the maximum running torque. The starting current is low and start-up will not cause a power consumption spike; it is also economical to run. The apparent efficiency is far higher than that obtained in a single-phase motor of conventional design. This motor is quiet and runs without brushes or commutators.

7.4.2 Three-Phase Squirrel-Cage Induction Motors

Squirrel-cage induction motors are fairly simple with the two shaft-supporting bearings and the rotor as the only moving parts. The stator is the stationary part that houses the electrical windings and the rotor is the rotating part that drives the load. In an induction motor, current is induced in the rotor by electromotive force (emf). There is no electrical connection from the power supply to the rotor, thus eliminating the need for brushes and commutators. The rotor is a cylindrical magnetic structure mounted on a shaft, with parallel slots in the surface. Bars are inserted in the slots and the bars are connected by a short circuit end rings on both sides. The name *squirrel cage* is derived from this structure. The current in the rotor conductors is induced by an electromotive force that is the result of the changing magnetic field around the stator winding.

As soon as electrical current moves through the stator windings, a rotating magnetic field moves across the stationary rotor conductors. This magnetic field induces an emf in the rotor conductors, causing current to flow in the rotor winding. This action is similar to the action of a transformer with the stator as primary winding and the rotor as secondary winding. The rotor conductors will cause their own force field, causing them to move at right angles to the magnetic field of the stator. The motor develops torque, causing it to accelerate in the same rotational direction as the rotating magnetic field in the stator.

It is interesting to note that the rotor cannot attain the speed of the magnetic field. If it could, then it would be static relative to the moving magnetic field. This would cause loss of induced electricity and thus emf. Since this will cause loss of torque, the rotor will slow down and because it will then be at a different speed, emf will be induced, torque will develop, and it will accelerate. The rotor speed will therefore always be less than the synchronous speed in an induction motor.

The difference between actual rotor speed (n_r) and synchronous speed (n_s) is called the slip. It is expressed as a percentage of the synchronous speed:

$$\text{Slip (\%)} = \frac{n_s - n_r}{n_s} \times 100$$

The average squirrel cage motor will operate at 2–3 percent slip. The synchronous speed of the motor is determined by the frequency of power system and the stator-winding configuration. The stator can be wound to produce two poles, one north and one south magnetic pole, or four, six, or eight poles. The synchronous speed (n rpm) is derived from frequency (f Hz) and the number of poles (p):

$$n = \frac{2f}{p} = \frac{2f\dfrac{\text{Hz}}{\text{s}} \times \dfrac{60\ \text{s}}{\text{min}}}{p} = \frac{120f}{p}\text{rpm}$$

Changing the number of poles can be used to control the speed in induction motors. The stator winding may be connected in such a way that one could use a switch to select the number of stator windings used. It is possible to connect the windings so that a twelve-pole motor running at 1200 rpm will run at 600 rpm when it is switched to six-pole operation.

It is also possible that stators can be provided with separate windings where one winding can provide for eight-pole operation and one for six-pole operation. This would yield synchronous speeds of 900 and 1200 rpm. Each winding can also be connected for a speed ratio of two to one by using an external switch, as discussed, thereby providing four speed selections. It must be pointed out that the speed steps are not continuously adjustable. Another way to alter the speed is to change the frequency of the power supply. This is possible with the use of the thyristor.

The torque of the motor is at a maximum when the rotor resistance and the rotor reactance are equal. The rotor resistance for a particular rotor is almost constant at all speeds but the rotor reactance will change with speed. Motors can be designed to have the desired torque characteristics at a specific speed.

To start a motor requires about six times the amount of current than that required running it at full load. For some large motors, this might be as much as nine times the full load current. This leads to very high consumption peaks with concomitant electrical surcharges. For this reason motors can be started at reduced voltage. The temporary reduced voltage will cause a lower starting torque, but it will limit the drop in line voltage that could damage other equipment. Another method that is used is to start the motor on a starter winding and then bring the other windings into action once the motor gets up to speed.

To change the direction of rotation in a three-phase motor, any two-line connections at the terminals need to be switched. Be very careful when using large motors that are not permanently wired. It has frequently happened that the terminals on the three-phase sockets are wired differently, causing some fairly horrible surprises when the motor is turned on.

7.5 Motor Management

In many processing plants, motors have to operate in high moisture or very dusty conditions. There are normally also wide load variations and operating conditions.

To get the best use out of a motor, it should be matched to the task that it needs to perform. When high starting torque is required, one should not use a motor with limiting starting voltage or current. The ventilation of the motor is also important. The casing is ribbed to increase the surface to help dissipate the heat. In very dusty conditions, these fins can be totally covered with dust or dirt, allowing the motor to run hot. Care should also be taken that the motor is suitable for wet areas.

It is important that motors are on circuits with the right size circuit breakers. For the purpose of maintenance, it is important to keep replacement motors in stock. It is much easier to work on a motor in a shop than to try fixing it on the line. The purchase cost will quickly be offset by time saved. It is also good policy that the mounting brackets for motors are kept standard. This will allow the use of a different size motor for short periods of time while maintenance and repairs are done to a motor in the processing line.

It has been found that the cause for burned-out motors will include one of the following:

1. A three-phase motor running on only one phase. (This is indicated by a loud humming noise.)
2. Improper fuse or overload relay set too high
3. Poor ventilation
4. Lack of lubrication
5. Low voltage
6. Overload
7. Excessive moisture
8. Ground or short in the motor

7.6 Power Transmission

Whenever power is transmitted or direction of movement is altered, loss of mechanical energy occurs. This loss due to friction and other phenomena causes only a fraction of the original power to be available at the point of use. Increased friction causes heat generation or increased thermal energy. That part of the mechanical energy that was used to convert into thermal energy accounts for loss of mechanical energy. This is the most important reason for the disappearance of the great steam locomotives. They were about 5 percent energy efficient; this means that 95 percent of the energy was not converted into traction. Most of the energy in the coal was lost as heat through inefficient heat transfer and energy conservation practices.

In a food processing plant, power is transmitted in many ways. For effective power transmission, the system used must be dependable, efficient, and as easy as possible to keep in a sanitary and safe condition.

7.6.1 Flat Belts

The flat belt used to be the most popular means of power transmission. On farms it is still used with the tractor as mobile power unit. Some older wheat mills also use a central drive shaft with pulleys and flat belts driving the different rollers in the mill. The flat belt, due to the smaller contact of surface area, was prone to slip.

Slippage causes a large loss in power transmission. Remember that any transmission that causes some part of the transmission system to heat up will convert some of the mechanical energy into heat; some of the mechanical energy is therefore lost.

Flat-belt drive systems are easy to maintain and relatively cheap. They are, however, difficult to fit with appropriate safety guards. Since the belts are also very long, it became common and rather questionable practice to splice broken belts by adding bits of older belt into the belt. This annulled proper preventative maintenance and caused many problems during processing.

7.6.2 V-Belts

V-belt drives became more popular and were the power transmission mode of choice for many years. They work very well and cause less energy loss. Multiple groove pulleys with increased surface area are frequently used and, by this means, large amounts of power can be efficiently transmitted. The poly-v-belts are, however, very dangerous and require good safety guards. They are also difficult to keep clean and hygienic and are frequently replaced by a direct shaft connection to a motor (see Figure 7.1).

A processing system requires fine adjustment of the speed of different processing lines to balance the flow of material. This requires speed adjustment capability for the different components, and variable speed motors and variable speed drives are used.

V-belts are quiet in operation and have great overload capacities. They also have considerable elasticity that helps to absorb shocks to a machine and driving system.

The load-carrying capacity of V-belts is determined by tests. Manufacturers rate belts and produce special tables that consider the size of belt, size of pulley, speed of

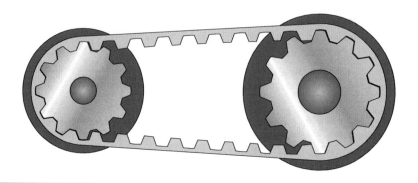

Figure 7.1 V-belt pulleys and belts.

belt, and type of belt. In addition, newer types of belts are available with a flexible metal cable incorporated, which increases the capacity and practically eliminates stretch. The cog V-belt is often used where no slippage can be allowed.

A variable speed drive can be made with variable diameter pulleys, which are very simple and trouble free. Since V-belts are not supposed to touch the bottom of the pulley, they have contact on the sides only. If the pulley is designed so that the sides can be moved closer or further apart, this will cause the belt to move closer or further from the axle, giving a variable diameter pulley. Some of these drives can be adjusted while running; for others, the sheaves must be adjusted and locked while stationary. The old VW Beetle used a variable diameter V-belt pulley to tension the V-belt.

Adjustable pulleys are widely used in processing systems. Speed can be changed from 25 to 50 percent with a system using one adjustable pulley. When more flexibility is required, systems are used that incorporate two sets of movable pulleys and a wide V-belt that can be moved without stopping. The speed can be adjusted and locked in position when conveyor speeds are matched.

The arc of contact on belts should be at least 120° for long wear and dependable service. Care consists essentially of maintaining good alignment of pulleys and proper tension on the belt, as recommended by the manufacturer. It is also important to keep the belts free from oil or water.

7.6.3 Gears

A direct shaft to the motor drive through a flexible coupling is the most efficient, quiet, and least expensive system of power transmission. It is not always feasible or possible to use a direct shaft as a coupling to the drive mechanism.

Gears are very widely used in the food processing industry because they are very efficient and long lasting if properly designed and made of suitable material. They also give exact control of speeds—for example, in the timing of processes, conveyors, etc.

Helical gears are normally quieter than straight spur gears, and hypoid gears are even more quiet and smoother in their action. Fiber or composition gears are very quiet if they are properly designed. Usually, a fiber pinion is used in connection with a gear train if extreme quietness is desired.

Cast iron gears are not used for heavy-duty service or for high-speed service. Usually, cut steel gear systems running in oil are used. In gear trains, gears of varying hardness are used and bronze gears are frequently used with steel gears. In all cases, a gear system should run in oil or have some lubricating system to keep it working properly.

In a closed gear system it is important to check the oil for water condensate buildup. The moisture is taken in due to "breathing" of air, since the atmospheric pressure changes from day to day and a surprising amount of condensate accumulates in the housing. Some manufacturers install a glass gauge on the gear housing to indicate when water has accumulated.

7.6.4 Calculation of Shaft RPM

It is important to be able to calculate the speed of pulleys, shafts, and gears. In a pulley system, the driving pulley is called the driver and the driven pulley the driven.

$$(\text{rpm} \times \text{diameter})_{\text{driver}} = (\text{rpm} \times \text{diameter})_{\text{driven}}$$

With this formula it is easy to calculate the rotational speed of the driven or driver. For gears and sprockets, the calculation is the same except that the number of teeth in the gear or sprocket is substituted for the diameter:

$$(\text{rpm} \times \text{teeth})_{\text{driver}} = (\text{rpm} \times \text{teeth})_{\text{driven}}$$

Example 7.1

A motor running at 5000 rpm has a pulley with a diameter of 15 cm. It is coupled with a V-belt to a machine that must run at 12,000 rpm. What must the diameter of the pulley on the machine be to get this speed?

$$\text{diameter}_{\text{driven}} = \frac{\text{rpm}_{\text{driver}} \times \text{diameter}_{\text{driver}}}{\text{rpm}_{\text{driven}}}$$

$$\text{diameter}_{\text{driven}} = \frac{5000 \text{ rpm} \times 15 \text{ cm}}{12000 \text{ rpm}} = 6.25 \text{ cm}$$

Changing the diameter to circumference and the circumference back to diameter is just multiplying and dividing with π.

7.6.5 Management of Drive Systems

Most drive problems are caused by misalignment of shafts, gears, sprockets, or pulleys; lack of lubrication; overloading; improper tension; or insufficient chain or pulley wrap.

A drive that is operating properly should be quiet and free from vibration. When replacing a chain or gear system, it is usually best to replace the driver, the driven, and the chain. Using old pieces with new pieces means continual maintenance that will cause a rapid increase in costs. The cost of labor is frequently more than the saving of old parts is worth. The loss of production caused by downtime is an even greater expense that is seldom added to repair costs.

A drive system that is also used in the processing industry is the hydraulic drive, in which oil under as much as 200 kPa is circulated by a special variable capacity pump and drives a hydraulic motor. Extreme ease of control of speeds over a wide range is possible with this system, and it is normally very quiet, safe, and sanitary.

Hydraulic controls that cover a wide range of uses are now possible with excellent control systems, such as are used in lift trucks and special operating units by which the operator can easily handle heavy loads by means of a small, easily operated control system.

7.7 Control Equipment for Electrical Apparatus

Many manufacturing industries and particularly food processors operate electrical equipment under the worst possible operating conditions, including corroding vapors, high moisture, and intermittent operation. The control equipment for electrical machines is frequently placed in remote locations away from the poor operating conditions. This could be a problem when one needs to switch off equipment in the case of an emergency and the switch is not really in an obvious place close to or on the machine.

In moist operating conditions, circuit breakers should be waterproof and operated by remote control. In the case of small motors, it is normal to use a waterproof circuit breaker and a waterproof push button to operate a magnetic switch. Most systems include a no-load release so that, if the switch is opened, they will not come on again until the push-button switch is pushed. This system will provide a low-voltage control since the switch will open if the voltage drops below a certain point.

Control equipment also includes overload relays that will open the circuit if the load becomes too heavy. Thermal overload relays are chosen according to the size of the load, which is determined by the motor size. These relays can normally be reset after about 1 min. A set of fuses will be used to back up overload relays. These fuses will burn out with continuously overloaded circuits.

7.8 Illumination of the Processing Facility

Pride in the workplace is easier to maintain in a well-illuminated plant than in a dark and dull facility. Working in an environment that is perceived to be *clean* promotes neat and tidy work habits. Good illumination thus enhances the operation of a well-run plant and promotes efficiency and safe working conditions.

Good lighting is an easy goal to reach and a quick fix to eliminate dark corners and unsafe work areas. The range of lighting hardware makes it possible to have a well-lighted plant. Industry recognizes some standards that should be met or exceeded.

When a lighting system is designed, the following points should be considered:

- The distribution pattern of the light and the suitability in the particular area involved
- Illumination output of the light hardware
- Possibility that larger lamps can be used in the same fitting when more light is required
- Design and construction of the lamp and its fitting
- The change in lamp efficiency over time and the ease of periodic servicing, cleaning, and replacement
- System cost

Visible light is limited to the spectral band from 380 to 770 nm. This is the part of the electromagnetic spectrum that is perceived by the human eye. Lamps will emit radiant energy in the visible and invisible spectra. The visible spectrum is composed of red, orange, yellow, green, blue, and violet, ranging from long to short wavelengths. The longer wavelengths, such as red and infrared, produce heat. The short wavelengths are more energetic and can influence reactions such as "arc eyes" when people work close to welding operations. It can also give the effects of sunburn. Most germicidal lamps emit light at 250 nm, whereas sun lamps peak at 300 nm and the "black light" that is used for inspection* peaks at 360 nm.

In the SI, light is measured in luminous intensity as candela. One candela (1 cd) is the luminous intensity, in the perpendicular direction, of a surface of 1/600,000 m^2 of a perfect radiator at the temperature of freezing platinum (1772°C) under pressure of one standard atmosphere (101.3 kPa). This is a very technical description for the old-style ft-candle. The lumen is the amount of light that will give a light intensity of 1 ft-candle over 1 ft^2 of surface, and this is the standard for the industry. It is further noted that 1 ft-candle is defined on the basis of using a standard candle at a distance of 1 ft from the source, while the lumen is the light of 1 standard candle over 1 ft^2 area at a distance of 1 ft. The foot-lambert is designated

---

* The presence of rodents in warehouses is easily seen by following the little pathways with "black lights." It is their dried urine that shows up so brilliantly.

as the measure of the brightness of illumination and is normally used as a measure of reflected light (10.76 ft-candles = 1 lumen m^{-2}). Remember that 1 m^2 is equal to 10.76 ft^2.

7.8.1 Light Intensity and Application

In any work area the light should be diffuse and uniformly constant. For the most efficient use of available light, the ceiling should have a minimum reflectance of 75 percent and the sidewalls 50 to 60 percent. The floor should be 20 percent reflective. To prevent eyestrain, glare should be avoided. The amount of light being reflected off any surface is affected by the smoothness of the surface. When the surface is rough, the reflection will be scattered and the reflected light will diffuse. When the paint surface is very smooth, irregularities in the painted surface can cause glare. The color of the paint will also affect the amount of light that is reflected (see Table 7.1).

Since walls are normally fairly smooth, color is the dominant factor to determine reflectance and illumination. Light colors reflect high proportions of the light, while dark colors absorb a lot of the light. Table 7.2 gives reflection values for different colors of paint. There is obvious variation between shades of the same color.

The human perception of a color is obviously influenced by the color of the light that illuminates it. When the dominant color in an area is cream, ivory, or tan, white fluorescent lighting will be best. If the dominant colors are blue or green, the blue type of fluorescent light will work best.

7.8.2 Types of Lamps

Many types of lamps are used in processing areas. In most cases, fluorescent lamps are favored since they have about 2.5 times the efficiency of incandescent lamps. They also give soft diffused light without glare. Fluorescent lamps are best suited

Table 7.1 Recommended Levels of Illumination

Offices	Illumination (Candela)
Designing, detailed drafting	200
Bookkeeping, auditing, tabulating, rough drafting	150
Regular office work, filing, index references, mail sorting	100
Reading or transcribing handwriting in ink or medium pencil	70
Reading high-contrast or well-printed material	30
Corridors, elevators, stairways	20

Table 7.2   Light Reflection by Different Colors of Paint

Color	Reflection (%)	Color	Reflection (%)
White gloss	84	Light blue	54
Flat white	82	Medium green	52
White, eggshell	81	Maple wood finish	42
Ivory white	79	Medium blue	35
Silver gray	75	Dark gray	30
Yellow	75	Oak wood finish	17
Cream	74	Walnut wood	16
Pink	72	Dark red	13
Light buff	70	Mahogany wood	12
Ivory tan	67	Dark brown	10
Medium yellow	65	Dark blue	8
Light green	65	Dark green	7
Medium buff	65	Black	5
Medium gray	58		

in areas where the lamp stays on for long periods of time. In places where lamps are frequently switched on and off, fluorescent light should not be used. Frequent on and off service not only causes a short life of the lighting element itself, but also places an extra load on the starting transformer. Fluorescent lamps can be used for about 2500 to 4000 h before they need to be replaced. Incandescent lamps must be replaced every 800 to 1000 h.

Most installations use fluorescent lighting in all the areas possible. There are some high-moisture areas where incandescent light fittings with vapor-proof fixtures are required. These areas include cold rooms and areas where explosive vapors may be present. In the cereal industry, the cereal dust can be very explosive when it is mixed with the right amount of air. In these cases, the light bulb and all the fittings are completely enclosed and watertight.

In areas where extensive coverage is required, mercury vapor lamps are used. Such areas include loading docks, large warehouses, and outside areas. Mercury vapor lamps are several times more efficient than fluorescent lighting.

Incandescent lamps radiate more long-wave radiation in the yellow and red range while fluorescent lighting is bluer. Incandescent lamps produce light and heat. This is an obvious drawback in cold storage areas. If fluorescent lighting is used in cold rooms, the tubes must be rated to operate at temperatures below 5°C.

The installation cost for fluorescent lighting is considerably greater than the cost for incandescent lighting. The energy savings will pay for this additional expense over time. All light bulbs should be replaced at regular intervals. Replacing them only if they are broken means that there will be one or two lights out at any given time. Light bulbs have an average life span and, for preventative maintenance, they should be replaced before they fail.

Chapter 8

Heating Systems for Processing Plants

8.1 Heat Transfer

Heat transfer is the movement of thermal energy from one point to another. The driving force behind heat transfer is the difference between the temperatures. Heat always flows from a high temperature to a lower temperature. Heat transfer is an equilibrium process that can take a considerable time to reach equilibrium.

Heat flow can be considered just like fluid flow, since we consider the rate of flow and the obstacles to flow. Thermal energy in a system manifests itself by the random motion of molecular particles within the system. The temperature of the system is just a physical manifestation of the particle motion. Heat transfer is thus concerned with the transfer of the molecular motion from one region to another.

Heat transfer occurs through conduction, convection, and radiation. In conduction, the energy is passed from one body or particle to an adjacent particle or body without any bulk movement of material. In convection, some bulk movement occurs and high-energy particles move to regions with lower energy. Convection takes place in liquids and in gases. Radiation is transfer of energy from a radiating source through space that can be void of matter.

8.2 Conductive Heat Transfer

Heat is transferred through solids by molecular excitation. Molecules that possess more energy vibrate faster than other molecules. When they collide with molecules that possess less energy, they transfer some of their energy to these molecules. In this way heat is transferred from the high-energy "hot" part of the system to the low energy "colder" part of the system. The heat transfer is because of the temperature difference.

Some materials, like metals, consist of atoms with large numbers of valence electrons that can vibrate. These materials are good conductors of heat because of the ease with which they can be excited. Materials in the nonmetal group have few valence electrons and are not very good conductors of heat. A poor heat conductor will act as a thermal insulator.

The closer together the molecules are packed in a system the better it will be for heat transfer. In gases, the molecules are far apart with relatively few interactions. Gases are poor thermal conductors. Liquids are better thermal conductors than gases since their molecules are packed together and can interact more readily.

Heat conduction is described as the time rate of heat flow ($\Delta Q/\Delta t$) in a material where a temperature difference (ΔT) exists. The rate of heat flow through a slab is directly proportional to the temperature difference (ΔT), and the area (A) of the slab, and inversely proportional to the thickness of the slab (d). Fourier determined that of heat transfer through a material is proportional to the temperature gradient.

$$\frac{Q}{t} \propto \frac{A\ T}{d}$$

With the use of a constant of proportionality, this can be converted to an equation:

$$\frac{Q}{A} = -k\left(\frac{dT}{dx}\right)$$

If the area A is constant over the length of the heat flow path and the thermal conductivity is considered to be constant, the formula can be integrated to

$$\frac{Q}{A} = -k\left(\frac{T_2 - T_1}{x}\right) = k\left(\frac{T_1 - T_2}{x}\right)$$

where Q (J/s) is the rate of heat transfer, k ($Jm^{-1}s^{-1}K^{-1}$ or $Wm^{-1}K^{-1}$) is the thermal conductivity, A (m^2) is the surface area, ΔT (°C or K) is the temperature difference, and x (m) is the thickness of the material. The minus sign on the thermal conductivity constant k ensures that the rate of heat transfer (Q) will be a positive number. Heat flows from a high temperature to a low temperature. T_1 will always be higher than T_2, thus making the difference ($T_2 - T_1$) always a negative number (see Table 8.1).

Table 8.1 Thermal Conductivity Values

Material	k (W/mK)	Material	k (W/mK)
Air	0.0242	Ice	2.2
Aluminum	240	Iron	47
Brick	0.71	Oxygen	0.024
Copper	390	Plasterboard	0.57
Corrugated cardboard	0.064	Stainless steel	21.5
Cork board	0.043	Styrofoam	0.042
Fiberboard	0.052	Water	0.571
Glass window	0.84	Wood, oak	0.15
Glass wool	0.042	Wood, pine	0.12

Example 8.1

A heavy-duty door handle on a freezer has a 16 cm square shaft of 1 cm × 1 cm that goes through the door. If the freezer is maintained at −20°C and the outside room temperature is +20°C, how much heat is transferred into the freezer by this shaft? Assume that the shaft is completely insulated along its length.

$$A = 0.01 \times 90.01 = 0.001 \text{ m}^2$$

$$x = 16 \text{ cm} \times \frac{1 \text{ m}}{100 \text{ cm}} = 0.16 \text{ m}$$

$$k = 47(\text{W/mK})$$

$$Q = \frac{kA(T_1 - T_2)}{x} = \frac{47(\text{W/mK}) \times 0.001 \text{ m}^2 \times (20 - (-20))/\text{K}}{0.16 \text{ m}}$$

$$Q = 11.75 \text{ W} = 11.75 \text{ J/s}$$

Multiplying this value with 60 × 60 = 3,600 will give the amount per energy entering the freezer per hour.

8.2.1 Resistance to Heat Flow

In many cases, the transfer of heat is not through a single homogenous layer but through series of layers where each layer has a different thermal conductivity value. In some of these layers the heat transfer rate will be slower than in others. The system of layers will have an overall heat transfer rate that is dictated by the thickness of the individual layers and their individual heat transfer rates. When the system

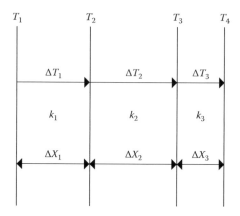

Figure 8.1 Diagram of the heat flow through a wall consisting of three layers.

has reached a steady state, the amount of heat transferred through the system will be the same as the amount of heat transferred through each individual layer. This concept is similar to a system of electrical resistors in series.

The Fourier equation integrates to give

$$\left(\frac{Q}{A}\right) x = k(T_1 - T_2)$$

This can be rearranged in the following form:

$$\frac{Q}{A} = \frac{kA(T_1 - T_2)}{x} = \frac{(T_1 - T_2)}{\dfrac{x}{k}} = \frac{(T_1 - T_2)}{R}$$

The thermal resistance (R) is equal to $\Delta x/kA$. The Fourier equation in this form is useful to simplify many calculations.

Look at the heat flow through a system with three different layers (Figure 8.1), where the temperature profile is described by T_1 to T_4. Under steady-state conditions, the heat flow through each of the three layers will be the same as the heat flow through each individual layer.

$$\frac{Q}{A} = \frac{(T_1 - T_2)}{\dfrac{x_1}{k_1}} = \frac{(T_1 - T_2)}{R_1} = \frac{(T_2 - T_3)}{R_2} = \frac{(T_3 - T_4)}{R_3} = \frac{(T_1 - T_4)}{\sum R}$$

This equation is the same as the calculation for current flowing through a series of resistors.

Example 8.2

The walls of a freezer consist of 1.5 mm stainless steel, 15 cm Styrofoam, and 3 mm aluminum. The thermal conductivity for the stainless steel is 21.5 W/mK, for Styrofoam it is 0.042 W/mK, and for aluminum it is 240 W/mK. The inside temperature of the freezer wall is –22°C and the outside temperature of the wall is 20°C. Calculate the heat transfer per square meter of wall.

$$R_{ss} = \frac{x_{ss}}{k_{ss}} = \frac{1.5 \text{ mm} \times \dfrac{1 \text{ m}}{1000 \text{ mm}}}{21.5 \text{ W/mK}} = 6.98 \times 10^{-5} \text{ m}^2 \text{ K/W}$$

$$R_S = \frac{x_S}{k_S A} = \frac{15 \text{ cm} \times \dfrac{1 \text{ m}}{100 \text{ cm}}}{0.042 \text{ W/mK}} = 3.57 \text{ m}^2 \text{ K/W}$$

$$R_A = \frac{x_A}{k_A} = \frac{3 \text{ mm} \times \dfrac{1 \text{ m}}{1000 \text{ mm}}}{240 \text{ W/mK}} = 1.25 \times 10^{-5} \text{ m}^2 \text{ K/W}$$

$$Q = \frac{(T_1 - T_4)}{\sum R} = \frac{(20 - (-22))\text{K}}{3.5700823 \text{ m}^2 \text{ K/W}} = 11.76 \text{ W} = 11.76 \text{ Js}^{-1} \text{ m}^{-2}$$

The heat flow is 11.76 J/sm² into the freezer. This is a simplistic solution since heat can also be conveyed by other means.

8.2.2 Heat Flow through Cylindrical Walls

When heat flows in a radial direction through a cylindrical body, there is a difference between the inside area and the outside area of the body. Area A of a cylindrical body of length L is given by

$$A = 2\pi r L$$

And the thickness of the wall is associated with the difference between the two radii. Substitution in Fourier's equation then yields

$$Q = \frac{-k(2\pi r L)dT}{dr}$$

In a steady-state situation, the transfer of heat does not change, so the formula can be rearranged to

$$\frac{dr}{r} = -kdT \frac{2\pi L}{Q}$$

This integrates between the limits of r_1 and r_2 to

$$\ln\left(\frac{r_2}{r_1}\right) = k(T_1 - T_2)\frac{2\pi L}{Q}$$

$$Q = \frac{2\pi L k(T_1 - T_2)}{\ln(r_2 / r_1)}$$

8.2.2.1 Steady-State Conduction through a Composite Cylinder

It is easier to work with the thermal resistance concept when working with composite cylinders. In this case, the thermal resistance of each layer is given by

$$R_1 = \frac{\ln(r_2 / r_1)}{k_1} \quad \text{and} \quad R_2 = \frac{\ln(r_3 / r_2)}{k_2} \quad \text{and so forth}$$

The overall heat transfer per area can then be written as

$$Q = \frac{2\pi L(T_1 - T_2)}{\sum R_i}$$

Example 8.3

A 5 cm standard steam pipe transfers steam at 130°C. The pipe is covered with a 2.5 cm thickness of glass fiber insulation (k = 0.042 W/mK). The outer temperature of the insulation is 40°C. What is the heat flow through a meter of pipe?

From Table 5.2 (Chapter 5), the inside diameter of a 50 mm pipe is 52.50 mm and the outside diameter is 60.33 mm. Assume the inside diameter of the insulation is 60.33 mm and the outside diameter is 60.33 + 2 × 25 mm = 110.33 mm.

$$R_1 = \frac{\ln(30.165 / 26.25)}{47 \text{ W/mK}} = 2.96 \times 10^{-3} \text{ mK/W}$$

$$R_2 = \frac{\ln(55.165 / 30.165)}{0.042 \text{ W/mK}} = 14.37253 \text{ mK/W}$$

$$\sum R = 14.3755 \text{ mK/W}$$

$$Q = \frac{2\pi \times 1 \text{ m} \times (130 - 40)\text{K}}{14.3755 \text{ mK/W}} = 39.34 \text{ W}$$

The heat flow through 1 m of pipe will be 39.34 W. If you want to calculate the temperature between the two layers, this can be done as follows:

Since the system is in a steady state, we have

$$Q = \frac{2\pi L k_1 (T_1 - T_2)}{\ln(r_2 / r_1)} = \frac{2\pi L k_2 (T_2 - T_3)}{\ln(r_3 / r_2)}$$

Determine the temperature between the two layers:

$$\frac{k_1 (T_1 - T_2)}{\ln(r_2 / r_1)} = \frac{k_2 (T_2 - T_3)}{\ln(r_3 / r_2)}$$

$$\frac{47 \text{ W/mK} \times (130 - T_2)\text{K}}{\ln(30.165 / 26.25)} = \frac{0.042 \text{ W/mK} \times (T_2 - 40)\text{K}}{\ln(55.165 / 30.165)}$$

$$0.604(6110 - 47T_2)\text{W/m}^2 = 0.139(0.042T_2 - 1.68)\text{W/m}^2$$

$$28.328T_2 = 3690.2$$

$$T_2 = 130.02°C$$

In this case, you can also work in kelvins and the temperature at the interface will be 403.02 K, which is the same as 130.02°C.

8.3 Convective Heat Transfer

Convective heat transfer involves the transfer energy from one molecule to another by conduction; the molecule then moves to a different part of the system, where it interacts with another molecule to release part of its energy. Convective heat transfer is therefore possible only in conditions where movement of molecules is possible, as in liquids, gases, and vapors.

Convective heat transfer will occur as a result of bulk movement in a fluid stream where a temperature gradient exists. Since conduction is an integral part of the convective heat transfer, the two means of heat transfer are not evaluated separately. Results of convective heat transfer automatically include the contribution of the associated conductive heat transfer.

There are two broad categories of convective heat transfer. One occurs naturally where the movement of fluid is driven by density gradients that result from temperature variations. In this case, thermal expansion of fluids is essential to drive the convection. Natural convection will occur in any large fluid-filled space whether it is desired or not. This is particularly important for gases and vapors where the viscosity of the fluid offers little resistance to flow.

The other category of convective heat transfer is forced convection induced by bulk movement of fluid through pressure gradients. In this case, there is direct control of fluid movement and the system can be designed to a required specification. Much larger fluid velocities are possible in forced convection with concomitant increase in heat transfer. In forced convection, it is essential also to determine whether the flow is turbulent or streamlined since this affects heat transfer.

Table 8.2   Values for Surface Heat Transfer Coefficients

Flow Situation	Convective Heat Transfer Coefficient BTU/(h ft² F)	Convective Heat Transfer Coefficient W/m²K
Still air	1	6
Moving air	5.3	30
Boiling liquids	420–10,560	2400–60,000
Condensing saturated steam	2113	12,000
Condensing steam with 3% air	616	3500
Liquid flowing in pipes: low viscosity	211–1056	1200–6000
Liquid flowing in pipes: high viscosity	21–211	120–1200

The rate of heat transfer by convection is proportional to the convective heat transfer coefficient, the area over which it takes place, and the temperature difference as a driving force. The rate of heat flow, according to Newton's law of cooling or heating, is proportional to the difference between the surface and fluid temperature:

$$Q = hA(T_s - T)$$

The proportionality constant, h, is the thermal conductance of the fluid film across which heat transfer takes place (see Table 8.2). The values of h will be the same as the values for k, but it also includes distance. To convert the heat transfer coefficients, one can use the following conversion factor:

$$1 \text{ BTU/hr} \cdot \text{ft}^2 \cdot {}^\circ\text{F} = 5.68 \text{ W/m}^2\text{K}$$

8.3.1 Heat Transfer by Conduction and Convection

Working with the both convective and conductive heat transfer opens up a whole new way to look at the system. A common heat transfer application involves the heat flow from a hot fluid through a solid wall to a colder fluid on the other side. In some cases, the wall can be flat but it is frequently tubular. There is also the presence of scale and other deposits that will cause even a simple system to become much more complicated.

To start, we will use the equation that describes conductive heat transfer:

$$\frac{Q}{A} = \frac{(T_1 - T_2)}{\dfrac{x}{k}} = \frac{(T_1 - T_2)}{R}$$

Add the convective heat transfer on either side:

$$\frac{Q}{A} = h_A(T_s - T) + \frac{(T_1 - T_2)}{R} + h_B(T_s - T)$$

The overall transfer coefficient U is defined as

$$\sum R = \frac{1}{h_A} + \frac{x_1}{k_1} + \cdots + \frac{x_i}{k_i} + \frac{1}{h_B}$$

$$U = \frac{1}{\sum R} = \frac{1}{\dfrac{1}{h_A} + \dfrac{x_1}{k_1} + \cdots + \dfrac{x_i}{k_i} + \dfrac{1}{h_B}}$$

$$Q = U(T_1 - T_2)$$

Example 8.4

You want to construct a continuous oven operating at 180°C. The oven will be installed in a room where the temperature is controlled at 20°C. To ensure a safe workplace, the outside wall temperature of the oven must always remain lower than 40°C. The convective heat transfer coefficients are 4 W/m²K inside the oven and because of air movement it is 2.5 W/m²K on the outside. The oven wall consists of two sheets of 2 mm stainless steel (k = 21.5 W/mK) with a glass wool layer (k = 0.042 W/mK) between them. What thickness of glass wool should be used to ensure that the oven temperature on the outside surface is lower than 40°C?

The heat flow through any part of the wall will be the same as the heat flow through the whole wall:

$$U_1 = \frac{1}{\dfrac{1}{2.5} + \dfrac{0.002}{21.5} + \dfrac{x}{0.042} + \dfrac{0.002}{21.5} + \dfrac{1}{4}}$$

$$U_1 = \frac{1}{0.4 + 9.3 \times 10^{-5} + 23.81x + 9.3 \times 10^{-5} + 0.25} = \frac{1}{0.650186 + 23.81x}$$

$$U_2 = \frac{1}{\dfrac{0.002}{21.5} + \dfrac{x}{0.042} + \dfrac{0.002}{21.5} + \dfrac{1}{4}} = \frac{1}{0.250186 + 23.81x}$$

$$Q = U_1(T_1 - T_2) = U_2(T_1 - T_s)$$

$$\frac{180-20}{0.650186+23.81x} = \frac{180-40}{0.250186+23.81x}$$

$x = 0.1071 \text{ m} = 10.71 \text{ cm}$

How much heat flows through a unit area of the oven?

$$Q = U(T_1 - T_2)$$

$$Q = \frac{(180-20)\text{K}}{0.650186+23.81\times0.1071 \text{ m}^2 \text{ K/W}} = \frac{160}{3.2} = 50 \text{ W/m}^2$$

This is a large amount of energy when one considers that an oven is about 100 m long by 0.75 m high and 1.2 m wide.

To calculate the heat transfer through pipes, you will use the following formulas:

$$R = \frac{1}{h_A A} + \frac{\ln(r_2 / r_1)}{2\pi L k_1} + \cdots + \frac{\ln(r_{i+1} / r_i)}{2\pi L k_i} + \frac{1}{h_B A}$$

All the other calculations are similsr to the previous example.

8.4 Heat Transfer by Radiation

All matter at a temperature above 0 K will continuously emit thermal radiation. Radiation has an associated wavelength and frequency, which will determine the character of the radiation. The product of the wavelength (λ) and the frequency (f) will give the velocity of the radiation. In a vacuum, it will be at the speed of light:

$$\lambda f = 3.0\times10^8 \text{ m/s}$$

If an object is heated, two things occur: The amount of energy that is released increases and the wavelength becomes shorter.

8.4.1 Stefan's Law

Stefan found that the amount of energy released from a heated object depends on the surface area, the temperature, and the nature of the surface. Surfaces that emitted the greatest amount of energy were black, so the name *black body* was given to surfaces with a standard energy distribution among wavelength and maximum energy at a specific temperature. True black bodies do not exist in nature. Stefan's law for black bodies is given by

Table 8.3 Emissivity Values

Material	Emissivity	Material	Emissivity
Stainless steel—smooth	0.2	Enamel fused on iron	0.9
Stainless steel—natural plate	0.44	Oil paint any color	0.94
Oxidized iron	0.7	Aluminum paint, 26% Al	0.3
Tinned plate	0.05	White paint	0.9
Aluminum foil	0.09	Water	0.95

$$Q = \sigma A e T_1^4$$

Q is the rate of emission in W, σ is the Stefan-Boltzmann constant (5.67×10^{-8} W/m²K⁴), A is the area in square meters, e is a unit less emissivity number that can be between zero and one that is characteristic of the material, and T_1 is the absolute temperature in kelvins. Dark surfaces have an emissivity close to one (see Table 8.3).

The rate of energy loss or gain by a body can be given by

$$Q = \sigma A e (T_s^4 - T^4)$$

If the answer is negative, there is an energy loss; there is an energy gain if the answer is positive.

Example 8.5

A 20 m insulated steam pipe with a diameter of 150 mm runs through a processing room. The temperature on the surface of the insulation is 28°C and the room temperature is 20°C. The insulation on the steam pipe is painted white (emissivity = 0.9). How much energy is radiated from the pipe in 24 h?

Since $(T_s^4 - T^4) \neq (T_s - T)^4$, we must first calculate the fourth powers of the absolute temperatures:

$$T^4 = (28 + 273)^4 = 8208541201 \text{ K}^4 = 8.21 \times 10^9$$

$$T_s^4 = (20 + 273)^4 = 7370050801 \text{ K}^4 = 7.37 \times 10^9$$

$$A = 2\pi r L = 150 \text{ mm} \times \frac{\text{m}}{1000 \text{ mm}} \times \pi \times 20 \text{ m} = 9.42 \text{ m}^2$$

Now all that is left is to plug the numbers into the formula and do a few calculations:

$$Q = \sigma Ae(T_s^4 - T^4)$$

$$Q = 5.67 \times 10^{-8} \text{ W/m}^2 \text{ K}^4 \times 9.42 \text{ m}^2 \times 0.9 \times (-838490400 \text{ K}^4)$$

$$Q = -4.03 \times 10^2 \text{ W} = -403 \text{ J/s}$$

The answer is negative and reflects heat loss. Convert the answer to reflect heat loss in 24 h:

$$Q = -403 \text{ J/s} \times \frac{60 \text{ s}}{\text{min}} \times \frac{60 \text{ min}}{\text{hour}} \times 24 = -34819200 \text{ J/h} = -34819.2 \text{ kWatt}$$

This system will cause a heat loss of 34.8 MW in 24 h.

What would happen if the insulation were covered with aluminum foil (emissivity = 0.09)?

$$Q = 5.67 \times 10^{-8} \text{ W/m}^2 \text{ K}^4 \times 9.42 \text{ m}^2 \times 0.09 \times (-838490400 \text{ K}^4)$$

$$Q = -40.3 \text{ W} = -40.3 \text{ J/s}$$

The emissivity factor change reflects directly in a smaller heat loss by a factor of 10.

8.5 Steam Heating

Many processes in industry require the use of steam as a heating medium. The operation of these plants and, especially, food processing plants is totally dependent upon a boiler with its associated system of steam distribution pipes. The heat processes that require steam in food industries include blanching, pasteurization, sterilization, canning, dehydration, and concentration. Some processes, like baking and drying, use other methods of heating.

To understand the energy value of steam, we need to understand isothermal energy changes that occur when the state or phase of a substance changes.

8.5.1 Enthalpy

Enthalpy is associated with the thermodynamic properties of a system. Enthalpy is a measure of the internal energy of the system under the influence of pressure and volume. In a case where the pressure is constant, isobaric, a change in enthalpy is equal to the heat transferred into or out of the system.

The thermodynamic definition of enthalpy (H) can be described as follows:

$$H = E + PV$$

where E = energy of the system, P = pressure of the system, and V = volume of the system. Enthalpy of a system is slightly different and is described as

$$H = mh$$

where m = the mass (kilograms) of the system and h = the specific enthalpy (kilojoules per kilogram) of the system.

This makes it relatively easy to calculate the enthalpy. In the metric steam tables given in Appendix 3, the value for the specific enthalpy is given for different temperatures. The subscripts for the specific enthalpy values refer to h_f for saturated liquid and h_g for saturated steam; h_{fg} refers to the difference between the enthalpy for saturated steam and saturated liquid.

Example 8.6

If you have 0.10 kg of water at 40°C in a container with a volume of 1.0 m³, what would the pressure in the container be? Describe the system.

Use the table for saturated water in Appendix 3.

At 40°C, the v_f = 0.001008 m³/kg and v_g = 19.52 m³/kg.

The average volume = V/m = 1.0 m³/0.1 kg = 10 m³/kg.

Because v_f is smaller than the average volume, which is smaller than v_g, the phase will be a liquid vapor mixture.

In the saturated state, the pressure at 40°C = 7.384 kPa.

The quality of the system is described by

$$x = \frac{m_{vapor}}{m_{total}} = \frac{v_{avg} - v_f}{v_g - v_f}$$

$$x = \frac{10 \text{ m}^3 \text{ kg}^{-1} - 0.001008 \text{ m}^3 \text{ kg}^{-1}}{19.52 \text{ m}^3 \text{ kg}^{-1} - 0.001008 \text{ m}^3 \text{ kg}^{-1}}$$

$$x = 0.512 = 51.2\%$$

8.5.2 *Energy*

In a thermodynamic system, the amount of transferable energy that a system possesses is called the enthalpy. When water undergoes a change of phase, the change in enthalpy is the latent heat of vaporization or condensation. When the temperature of a body changes, the enthalpy changes with every degree change. Enthalpy is the heat capacity of the system at constant pressure. The heat content or enthalpy of a body tends to go from a higher to a lower state of energy. Heat will therefore flow from the warmer system to a colder system.

Entropy is somewhat more difficult to describe. It is a natural phenomenon that systems will move from an orderly state to a disorderly state.* The natural tendency is that entropy will increase while the available energy decreases.

---

* Farming is a system where we remove the diverse plants and change to a single crop for that season. When the land is left dormant, the normal diversity will return.

8.5.3 Specific Heat Capacity

The specific heat capacity of a substance (J/kg K) is the amount of energy required to raise the temperature of 1 kg of the substance through 1 K. (Remember that a difference of 1 K = 1°C). To raise the temperature of m kg substance, with a specific heat capacity of c, T K we need

$$Q = (mcT) \text{joule}$$

The specific heat capacity of water is 4.186 J/(kg K).

8.5.4 Specific Latent Heat

The specific latent heat (joules per kilogram) is the energy required to change the phase of 1 kg of substance without a change in temperature. When the change is from solid to liquid, it is the specific latent heat of fusion and when it is from liquid to vapor, it is the specific latent heat of vaporization. The heat (Q) required to change the state of m kg of substance of specific latent heat L J/kg is

$$Q = (mL) \text{joule}$$

The specific latent heat of water is 2252.4 kJ/kg.

8.5.5 Energy Associated with Steam

The distribution of heat in the plant and the application of heat to the product require careful analysis and planning to ensure high efficiency. Steam is a very good heating agent because it carries a great amount of heat per kilogram. At atmospheric pressure, 1 kg of steam will release 2252.4 kJ (2134.9 Btu) when it condenses.

When energy is added to water, the temperature of the water will increase until it reaches boiling point. When more energy is added, the water will change into vapor. The added energy in the form of heat caused the vaporization of the water and allowed the vapor to escape from the surface of the water when the vapor pressure inside the water exceeded the atmospheric pressure above the water surface.

Dry steam is a colorless vapor that becomes visible as a whitish cloud when it is wet. Steam is lighter than air and tends to rise, causing it to condense on any structure that is cold enough to absorb the energy. This normally occurs on the ceiling, where the continuous wet areas give rise to mold growth. The spores from the mold will contaminate everything and are a terrible nuisance. Many plants go to great lengths to solve the problem and in many cases the solution is the seasonal replacement of the ceiling.

Vaporization of water causes it to undergo a very large change in volume. A liter (kilogram) of water at atmospheric pressure (101.3 kPa) will occupy a volume of 1.65 m³ at 101.3 kPa. At increased pressure of 344.7 kPa, it will occupy 0.19 m³.

8.6 Heat Pressure Diagram for Water

The properties of steam are best explained when one considers the energy used to change water into vapor. To heat 1 L of water (1 kg) at 0°C to 100°C requires 418.7 kJ. The heat that is absorbed is sensible heat since it causes a change in temperature.

The 1 kg of water will stay at 100°C while it absorbs more energy. The temperature will stay constant until another 2257.15 kJ of energy has been absorbed. The heat that was absorbed is called latent heat since there was no change in temperature with a large change in total energy.

At this point, the water absorbed a total of 2675.85 kJ/kg. We need to add only 2 kJ of energy to increase the temperature 1°C beyond 100°C. The steam, which is now at a temperature greater than 100°C, is called, superheated (see Figure 8.2).

When this 1 kg of saturated steam condenses back to water, all of the 2257.15 kJ of energy will be transferred to the condensing surface. This property makes steam a particularly good heating medium. The values given before are for atmospheric pressure. All the values, including temperature, will change when there is a change in pressure. The relationship of pressure, temperature, specific volume, latent heat, and others have been calculated and are given in the steam tables in Appendix 3.

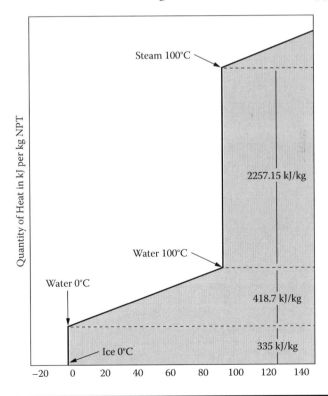

Figure 8.2   Heat temperature diagram for water.

It should be noted that the steam table is usually based upon zero heat at 0°C and gives the heat above this point. A separate table is used for superheated steam.

Example 8.7

An electric kettle was half full of water when it was switched on and allowed to boil. While the water was boiling steadily at 100°C, 2 L of water at 20°C was added to the kettle. It took 5 min before the water was boiling again.

1. What is the power of the heating element in the kettle? The specific heat capacity for water is 4.187 kJ/(kg K).
2. How much water will vaporize when the kettle is left to boil for 10 min? The specific latent heat of vaporization of water is 2257.15 kJ/kg.

The water that was in the kettle was brought back to the same temperature that it had been before the cold water was added, so all the energy was used to heat the 2 L of water through 80°C.

$$Q = mcT = \left(2 \text{ liter} \times \frac{1000 \text{ cm}^3}{\text{liter}} \times \frac{1 \text{ g}}{\text{cm}^3} \times \frac{1 \text{ kg}}{1000 \text{ g}} \right) \times \frac{4.187 \text{ kJ}}{\text{kg K}} \times 80 \text{ K} = 669.9 \text{ kJ}$$

$$\text{Power} = \frac{669.9 \text{ kJ}}{5 \text{ min}} \times \frac{1 \text{ min}}{60 \text{ s}} = 2.233 \text{ kJ/s} = 2.233 \text{ kW}$$

Since we know the time and the specific latent heat of water, we can solve for the mass:

$$m = \frac{Q \text{ kJ/s}}{L \text{ kJ/kg}} = \frac{Q \text{ kJ}}{\text{s}} \times \frac{\text{kg}}{L \text{ kJ}}$$

$$m = \frac{2.233 \text{ kJ}}{\text{s}} \times \frac{10 \text{ min} \times 60 \text{ s}}{\text{min}} \times \frac{1 \text{ kg}}{2257.15 \text{ kJ}} = 0.59 \text{ kg}$$

8.6.1 Properties of Steam

Water can exist as saturated liquid or vapor (steam) when it is above its freezing point. Steam can exist as saturated vapor, vapor–liquid mixtures, and superheated steam. All three forms are important in the food processing industry.

8.6.1.1 Saturated Liquid

A liquid with a free surface in contact with a confined volume of air will equilibrate with the air such that the air becomes saturated with vapor. The vapor will exert a saturated vapor pressure on the liquid surface. If the vapor pressure above the liquid exceeds vapor pressure, then other gases are present in the atmosphere above the liquid. When the pressure above the liquid is equal to the vapor pressure for the liquid at that temperature, the liquid will be at its boiling point.

This leads to the interesting phenomenon that water can boil at any temperature between 0°C and 100°C depending on the pressure of the atmosphere directly above it. In fact, with enough pressure, we can have water not boiling at 2000°C. Water at this temperature is used to dissolve substances to make artificial emeralds.

8.6.1.2 Saturated Vapor

Saturated vapor is steam, in which all the water is vaporized, at the boiling point of water. Even very slight cooling at constant pressure will result in condensation. The change in phase will cause a release of heat. If the heat is removed from the system at constant pressure, all the steam can condense before a change in temperature is observed. The addition of more heat will cause superheating, which will cause an increase in temperature or pressure or both. For every pressure, there is a corresponding temperature. If a higher temperature is required, the system should be designed to handle the concomitant higher pressure.

8.6.1.3 Wet Steam

Steam can contain small droplets of water in it. This water has not undergone the phase change and therefore does not have the additional energy associated with it. The wet steam therefore has less than the optimal amount of energy associated with it. The steam quality is given as the percentage of a vapor–liquid mixture that is in the form of saturated vapor. For example, if 1 kg of 95 percent quality steam is used at zero gauge pressure for heating, it will give up only 0.95×2257.15 kJ, or 2144.3 kJ, upon condensing. The total heat of the wet steam is latent heat plus sensible heat.

There are many disadvantage associated with wet steam. In regular use it causes abnormally high amounts of condensate in the system. If the system was not designed to handle excessive condensate, this can lead to condensate carryover in the steam lines. If steam turbines are used, the wet vapor will abrade the turbine blades much more rapidly than saturated steam. The first indication of wet steam or "carryover" is when the boiler feed pump is working more than normally. One gets used to the regular patterns in the boiler operation and any change will be noticed. This will be followed by Production complaining about lack of steam.

Wet steam can be caused by any of the following conditions:

1. High solid concentrations build up inside the boiler due to inadequate blowdown.
2. The feedwater was not treated right or the treatment was inadequate.
3. The steam line at the boiler nozzle is undersized, causing excessive steam velocities, which entrains water out of the boiler.

4. Quick opening steam valves cause surges in load, thereby causing instantaneous boiler overloading.
5. Plant expansion causes constant boiler overload.
6. The steam header is waterlogged because of a lack of steam traps.

8.6.1.4 Superheated Steam

When steam contains more energy than what is required to vaporize all the moisture fully, it is superheated. The number of degrees in excess of the vaporization temperature is the degrees of superheat. Increasing the temperature of saturated vapor produces superheated steam. Passing dry saturated steam through specially heated coils on its way to the steam system is the way that it is usually produced. These superheating coils are usually located in the top of the combustion space of the boiler.

The temperature will rise when heat is added to steam to produce superheated steam. At atmospheric pressure, 1093 J (0.47 Btu/lb) are absorbed per kilogram of steam, per degree of temperature increase.

Since temperature and pressure are related, passing steam at high pressure through a pressure-reducing valve to a state of lower pressure can also produce superheated steam. At this lower pressure, the steam will contain more heat than that required for vaporization. Superheat is advisable when very long distribution pipelines are used. This will reduce the amount of condensate that has to be returned to the boiler.

8.7 Steam Tables

When the vapor pressure of a liquid is equal to the external pressure on the liquid, movement of molecules between the liquid and vapor phase is enhanced. This rapid vapor transfer state is normally called boiling. The relationship between boiling point and pressure is used for many calculations where it is important to know how much vapor can be in the atmosphere above a liquid.

For a table like the metric steam table in Appendix 3, a base state is chosen, and in this example the enthalpy and entropy of the saturated liquid are equal to 0.0 at a temperature of 0°C.

The first two columns in the steam table show the relationship between temperature and saturated vapor pressure. It indicates how the boiling point of water changes with pressure. The figures can be used to determine the boiling point of water at any given pressure. The absolute pressure at any given point is also the vapor pressure.

The third column indicates the specific volume of the vapor and is equivalent to the reciprocal of density; it is given in cubic meters per kilogram. This column shows the relationship between the volume of a unit mass of saturated vapor and the pressure.

Enthalpy is an indication of the heat content of 1 kg of steam or water at the temperature and pressure indicated. It would be more accurate to refer to it as specific enthalpy since it is based on a unit value for the mass. The enthalpy of moist vapor consists of the enthalpy of the saturated vapor (subscript g) and the enthalpy of the saturated liquid (subscript f). The difference between the enthalpy of the saturated vapor h_g and the enthalpy of the saturated liquid h_f is given in the column h_{fg} and this value represents the latent heat value at the given operating conditions.

Changes in entropy take place when evaporation or condensation occurs. The entropy changes associated with these processes are given by the value in column S_{fg}, where

$$S_{fg} = S_f - S_g$$

Since phase changes occur at constant pressure, the following relationship can be applied:

$$S_{fg} = \frac{h_{fg}}{T}$$

Taking any value in the steam table, it can be seen that at 90°C, the specific enthalpy will be 2283.2 kJ/kg.

The entropy at 90°C + 273 = 363 K can be calculated as

$$S_{fg} = \frac{h_{fg}}{T} = \frac{2283.2 \text{ kJ}}{\text{kg}} \times \frac{1}{363 \text{ K}} = 6.29 \text{ kJ/kg K}$$

It is useful for anyone working with steam to be familiar with the steam tables since they describe the relationships between temperature, pressure, and heat. They are also the tool for the solution of many heating problems.

The imperial steam table in Appendix 3 starts with steam at atmospheric pressure (0 gauge pressure) and temperature of 212°F. Other tables are used for lower temperature and pressure relationships. From the steam table, one can determine what pressure of steam will be required to produce a given temperature and also calculate the amount of steam needed to provide a given amount of heat in BTUs.

The specific heat of dry saturated steam increases with pressure, and the specific heat is higher than for air. This is why steam should be pure and not mixed with air. Latent heat per kilogram of steam decreases with increase in temperature of evaporation and there is a total heat increase with increase of temperature of evaporation and an increase in superheat.

8.8 Boiling Point of Water

Since boiling occurs when the vapor pressure is equal to the pressure upon the surface, it follows that altering the pressure upon the water will alter the boiling temperature. There is a temperature of evaporation that corresponds for each pressure. This temperature–pressure relationship is of great practical importance in steam heating operations since the temperatures of processing utilizing saturated steam can be controlled by means of pressure control.

8.8.1 Heating with Live Steam

Steam as a heat transfer medium is used in many systems. In its simplest form, the product is placed in an enclosed chamber and the steam is introduced into the space surrounding the product. This process is the basis for applications such as steam peeling, steam blanching, and many others. There are some guidelines that must be followed.

Steam under pressure will fill all the space and thereby ensure an even and uniform temperature. As the steam particles condense, the phase change will release 2257.15 kJ energy per kilogram of steam at atmospheric pressure. At different pressures, the temperatures will be different and the steam will maintain a certain temperature for every pressure.

The condensate must continuously be removed or else it will eventually fill up the container. It is good to remember that once the product is submerged, the condensing steam will release its energy to the water and not to the condensing surface.

8.8.2 Culinary Steam

Boiler water is frequently treated to optimize the performance of the boiler. The feedwater can contain chemicals that are not intended to contaminate food. Even when no treatment of the water is done, the boiler water can concentrate the dissolved and colloidal material that was present in the water through continual loss of pure water vapor. Some of the boiler water will find its way into the steam as atomized liquid droplets. This wet steam coming from the boiler can contain impurities that will contaminate the food if it is allowed to come into contact with it. To ensure the well-being of the consumer, food law requires that culinary steam be used when it is injected in food products (see Figure 8.3).

Passing steam that is produced by a regular boiler through tubes in a secondary boiler produces the culinary steam. The steam becomes the heating source to produce the steam in the secondary system. The secondary boiler produces the culinary steam from potable water. This steam will be uncontaminated. The primary steam is frequently superheated to ensure efficient steam generation in the secondary boiler. Superheating is done with a heat exchanger.

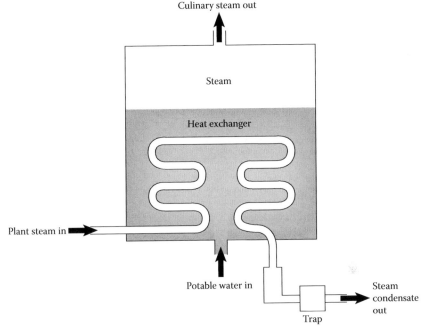

Figure 8.3 Culinary steam boiler.

8.9 Hot Water Systems

Many operations rely on water at a specific temperature. In the cleaning operations, large quantities of hot and warm water are used to clean away the deposits of the production shifts. Hot water heaters are very important in food plants and are sometimes an incredible cause of wasted energy. In large plants, hot water heaters and storage tanks are used as the most efficient way to control the temperature and the cost.

In some cases, the use of piped water is not feasible and on the spot heating is accomplished by a mixing valve. The McDaniel Tee is frequently used for heating of cleaning water. The valves can waste large quantities of steam if they are not properly shut off after use (see Figure 8.4).

8.9.1 Energy Calculations for Water Heating

Energy must be added to water to heat it. The amount of energy required is given by the formula

$$Q = mc\Delta T$$

where m = mass (kilograms) of water to be heated, c = specific heat of water (4186 J/kg·°C), and ΔT is the temperature difference.

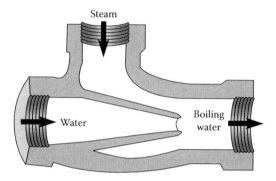

Figure 8.4   Heating valve.

Example 8.8

How much energy is required to raise the temperature of 2000 kg water from 18°C to 67°C?

$$Q = 2000 \text{ kg} \times \frac{4186 \text{ J}}{\text{kg °C}} \times (67 - 18)°C$$

$$Q = 410228000 \text{ J}$$

$$Q = 410.2 \text{ MJ}$$

Since heat transfer from one system to another is frequently associated with some loss, the efficiency of the heat transfer should be considered. The formula becomes

$$Q = \frac{mc \ T}{eff}$$

If the transfer efficiency is 0.89, how much heat will be required for the problem in Example 8.7?

$$Q = 2000 \text{ kg} \times \frac{4186 \text{ J}}{\text{kg °C}} \times (67 - 18)°C \times \frac{1}{0.89}$$

$$Q = 460.9 \text{ MJ}$$

8.10 Piping

Piping is a considerable expense in processing plants. For steam, the pipe used must be designated steam piping of sufficient strength to cope with the temperature and pressure loads. Prior to initial operation, the pipe system will have to be tested for any leaks in the system. There are different air and hydraulic pressure tests that are performed to ensure that the system can cope with the design pressures.

For the final test, a hydrostatic pressure test is done. The hydraulic pressure used in the test shall be greater than 1½ times the design pressure. If the design temperature is greater than the test temperature, the minimum test pressure is calculated using the following formula:

$$P_T = 1.5P\frac{S_T}{S}$$

where P_T is hydrostatic gauge pressure (MPa), P is the internal design pressure (MPa), S_T is allowable stress at the test temperature (MPa), and S is the allowable stress at design temperature.

8.10.1 Expansion of Pipes

Heating and cooling of pipes cause expansion and contraction. These changes in dimensions are common thermal effects that can cause large problems. Thermal expansion results from the change in the average distance between adjacent atoms in the solid or liquid. A change in one direction in a solid is called linear expansion. The change in length $(L_2 - L_1)$ is somewhat proportional to the change in temperature $(T_2 - T_1)$. We use a thermal coefficient of linear expansion (α) to give

$$L = \alpha L_1 \ T$$

$$(L_2 - L_1) = \alpha L_1 (T_2 - T_1)$$

$$L_2 = L_1 + \alpha L_1 (T_2 - T_1) = L_1 (1 + \alpha[T_2 - T_1])$$

Example 8.9

There is a steam line running from the steam plant to a process. The line runs along the outside wall of the factory. The normal operating temperature of the line is 130°C. When the line was installed at 20°C, it was 100 m long. How long will it be in operation? The thermal expansion for steel is $12 \times 10^{-6}/°C$.

$$L_2 = 100 \text{ m} + 12 \times 10^{-6} \ /°C \times 100 \text{ m} \times (130 - 20)°C$$

$$L_2 = 100.132 \text{ m}$$

The steam line will be 132 mm longer. When it cools down, it will be shorter and this can put strain on the pipe fittings.

To deal with expansion in long pipelines, we normally use expansion or slip joints, or we install a bend in the pipe. For steam pipes it is important that the condensed water will run back to the boiler or to a trap. With the installation of bends in the line, this must be kept in mind; otherwise, little traps full of condensate will result.

Problems

1. An oven has a 60 mm thick wall consisting of 2 mm aluminum on the outside and 1.5 mm stainless steel on the inside. The cavity between the metal sheets is filled with glass wool. If the inner surface of the oven is at 200°C and the outside surface is at 22°C, calculate the amount of heat that is transferred through a square meter of the oven wall.

2. A room for accelerated shelf-life studies is maintained at 35°C. The temperature outside the room is maintained at 20°C. Electricity is used to heat the room at a cost of $0.08 per kilowatt. The room is 3 m wide, 4 m long, and 2 m high. The walls and the roof of the room are made from the same material. The contractor submits two quotes. In one they will construct the room with 20 mm plasterboard siding and 100 mm Styrofoam as insulation at a cost of $20,000. In the other quote, they will use 18 mm pinewood on either side with 150 mm Styrofoam as insulation at a cost of $30,000. Calculate the difference in heating cost over a year and decide which quote to accept. Assume that the convective heat transfer coefficient is 4 W/m²K for all surfaces.

3. A manufacturer wants to put a glass window in an oven. To ensure the safety of the household, the oven window should never have a temperature of more than 50°C. To clean the oven, a temperature of 225°C is used. Calculate the air gap (k = 0.0242 W/mK) between two sheets of 5 mm glass (k = 0.84 W/mK). The convective heat transfer coefficient on both sides of the glass is 6 W/m²K.

4. Milk flows through a 25 mm inside diameter stainless tube that is heated from the outside by steam. The experimental data show initial overall heat transfer (U) of 1700 W/m²K. After some time, a deposit inside the tube caused the overall heat transfer to reduce to 250 W/m²K. What is the value of the heat transfer coefficient of the deposit? These are all the data needed; nothing was omitted.

5. A stainless steel tube having an outside diameter of 32 mm and a wall thickness of 1 mm is going to be insulated with a material with a thermal conductivity of 0.042 W/mK. If both the inner and outer convective heat transfer coefficients are 6 W/m²K, what is the heat loss in 1 m of pipe if the inside temperature is 95°C and the outside temperature is 20°C? Use insulation thickness of 10, 20, 30, and 40 mm. Draw a graph of heat loss per meter against insulation thickness.

6. In an area with very hard water, scale buildup is a problem. The factory maintenance manager decided to check on the amount of scale buildup inside a steel pipe carrying hot water at 100°C. The water flow is turbulent and the

surface temperature inside the pipe can be assumed to be 100°C. The pipe is insulated with 25 mm of glass wool. The pipe has an outside diameter of 52 mm and an inside diameter of 50 mm. The room temperature is 20°C and the convective heat transfer coefficient on the outside of the insulation is 2.5 W/m²K. What will the temperature on the insulation surface be just after the pipe was installed and once a 3 mm thick calcium deposit has formed? The calcium deposit has heat transfer coefficient of 0.56 W/mK.

7. After sterilization of bright tin cans, the cans are removed from the retort at a surface temperature of 80°C. They are placed in an area where the room temperature is 20°C. How much heat is lost through radiation per minute? Do the same calculation for surface temperatures of 70°C, 60°C, 50°C, 40°C, and 30°C. What conclusion can be drawn regarding emissivity?

Chapter 9

Steam Generation

Within the food industry, steam is used for processing, drying, and heating and for general use such as in sanitizing. Some type of steam generation system is included in most food processing plants. The demand for steam varies in different operations and in some plants or parts of the plant, steam is used continuously while it is used intermittently in others. The steam generation system must be adapted to the operations that will be performed within a plant. Since the boiler room is the energy source for most processes, the system needs to be designed with possible expansion and maintenance in mind.

9.1 Boilers

Steam is produced in a boiler or steam generator. The boiler is housed in a building that is frequently separate from the main processing area. It is always surprising to step into a boiler room to find the cleanest and best maintained piece of equipment on the premises. There are many types and sizes of boilers that can be found in industry.

In very simple terms, a boiler can be seen as a closed container with water under pressure in it. In a furnace the chemical energy of a fuel is converted to heat energy that must be transferred to the water in such a way that as little as possible of the heat is lost. The heat of combustion is transferred to the water with conduction, convection, and radiation. The contribution of each of these modes of heat transfer depends upon the type and design of the boiler.

The two principal types of boilers used for industrial applications are fire tube and water tube boilers.

9.2 Fire Tube Boilers

In a fire tube boiler (Figure 9.1), hot combustion gases run through the boiler tubes. The tubes are surrounded with the water that will be heated. The tube system is contained in a large pressure vessel that will also contain the water and steam. In some applications the furnace is also housed within the same vessel. Fire tube boilers are available with pressure outputs of 103.4 to 2413 kPa. When higher pressures are required, the shell of the boiler needs to be made of thicker material. They are available to produce steam in a range from 10 to 1100 kW.

Fire tube boilers are designed to work with large water content and require a large space for the water pressure vessel. The boiler requires a large amount of floor space. The boilers take a long time to produce steam from a cold start and are therefore not suitable for intermittent use. Maintenance of the firebrick, refractory system, and of the boiler is required every 3 – 9 years.

Fire tube boilers are, however, very reliable and capable of producing high-quality stream. They have a large water capacity and can handle variable loads with almost no variation in steam pressure. A fire tube boiler can be envisaged as a shell and tube heat exchanger. The large number of tubes causes a very large surface area, resulting in a large heat transfer area. This increases the overall boiler efficiency.

Combustion of the fuel occurs in a refractory lines furnace and the flue gases are circulated through the tubes to the stack outlet. When the water absorbs most of the heat, very little of the heat will be lost in the stack, resulting in high boiler efficiency. The hot gases can pass through the boiler in a number of ways. In a single-pass design, the gas passes through the tubes and out through the stack. In a four-pass design, the hot gases are rerouted through a second, third, and fourth set of tubes.

A boiler with **four** gas passes is **normally *more*** efficient than one with fewer gas passes. The main aim is to remove as much heat as possible from the gasses before they exit into the stack. The methods include the use of baffles to create more turbulence, the use of economizers, and an increase in the tube surface area.

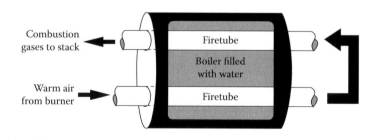

Figure 9.1   Fire tube boiler.

9.3 Water Tube Boilers

In water tube boilers (Figure 9.2), the combustion gases pass around tubes filled with water. The tubes are interconnected on the feed side and the steam side. The water tube boiler is often called a safety boiler since there is less chance of a serious accident when one of the water-filled pipes inside the boiler bursts. The explosion would be a nonevent compared to what would happen if the large drum of a fire tube boiler failed.

Water tube boilers can normally operate at higher pressures than fire tube boilers. In power plant applications, pressures of 10,342 to 13,789 kPa are often used.

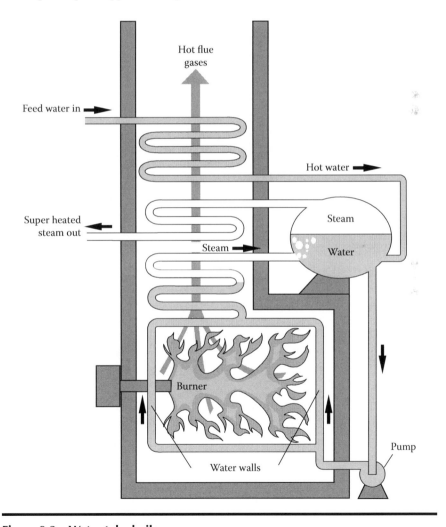

Figure 9.2 Water tube boiler.

The water tubes contain steam and steam and water mixtures while the hot combustion products pass around the tubes.

Water tube boilers can have more than one drum. Tubes, some of which can be in the form of waterfall tubes lining the furnace area, connect the steam drum and mud drum. Water tube boilers are rated in kilograms of steam output per hour and they can deliver from 4,500 to 61,000 kg/h. The boilers are faster than fire tube boilers to come up to full pressure since they contain a relatively small amount of water. They can respond quickly to load variations.

The design of water tube boilers allows for the production of super-heated steam. In the case of load variation and where high pressure or superheated steam is required in the process, the use of a water tube boiler should be considered.

9.4 Boiler Efficiency

Efficiency is a measure of the economic use of resources. In the case of a boiler, the efficiency is a way to measure how well the chemical energy in some fuel can be converted to heat in steam. There are different ways to define efficiency and different interpretations of efficiency. Obviously, there can thus be a lot of confusion about efficiency.

Boiler efficiency and fuel usage are two of the more important selection criteria when a boiler is considered. Boiler efficiency is a calculation of the difference between the energy input and energy output. Every boiler operates under the same fundamental thermodynamic principles. Therefore, a maximum theoretical efficiency can be calculated for a given boiler design. The maximum value represents the highest available efficiency of the unit.

9.4.1 Fuel-to-Steam Efficiency

Fuel-to-steam efficiency is the most acceptable efficiency rating used for boilers. There are two ways to calculate this efficiency according to ASME Power Test Code. The first way is to calculate the percentage ratio of energy output divided by energy input. The second method is to calculate the amount of energy loss through radiation and convection as well as loss in the flue and then subtract it from the total energy. The values are converted so that total energy input equals 100.

9.4.1.1 Combustion Efficiency

Combustion efficiency gives an indication of the effectiveness of the burner. It describes the ability of the burner to burn the fuel such that all the chemical energy is released. Combustion is the first step that dictates the overall efficiency. Any problems with burner efficiency will reflect in overall efficiency. A burner of good design that is adjusted for optimal performance will be able to give complete combustion with only 15 to 20 percent excess air.

$$CHOS + O_2 = CO_2 + H_2O + SO_2$$

The amount of each combustion product formed will depend on the amounts of the atoms in the fuel. The composition of the fuel will also dictate the oxygen requirement. Any lack in oxygen will obviously cause the formation of less oxidized combustion products, leading to loss of energy in the flue gases. For this reason excess air is used to provide a safe oxygen excess. When high excess air is used, perfect combustion is assured, but the heat energy will be wasted on heating the excess air rather than the water. The high volume of air will also reduce residence time of the hot gases inside the heat transfer areas and thereby increase the stack loss.

It is very important to determine what the excess oxygen requirement for different boiler loads will be. Having a standard amount of air blown into the burner will lead to excessive air in the burner when the boiler is running at low loads of perhaps 20 or 30 percent of full load. Efficiency calculations should be made for different loads so that accurate costing of heat units can be made.

Before any decision is made regarding preheating of intake air, one should balance the additional equipment cost against possible increases in burner efficiency. Frequently, it just is not a viable option.

9.4.1.2 Thermal Efficiency

Once the efficiency of the burner has been optimized, the efficiency of the boiler needs attention. Boiler efficiency is a function of design and one can improve the performance to be optimal within the design constraints. All the heat transfer surfaces must be kept in a condition that will promote heat transfer rather than hinder it. True boiler efficiency is a combination of fuel and thermal efficiency.

Preheating the makeup water can increase boiler efficiency. It is common practice to return the hot condensate to the boiler as makeup water. Makeup water can also be preheated in the stack at a very small additional equipment cost.

9.4.1.3 Stack Temperature

A high stack temperature is a sure sign that a large amount of heat is being vented out of the boiler with the combustion gases. In an efficient boiler, most of the heat would have been transferred to the water and only a limited amount of heat would be lost in the stack. Monitoring the stack temperature gives a good indication of any heat transfer problems in the boiler. When the stack temperature rises continuously, it is perhaps an indication that the heat transfer surfaces are in need of maintenance. The amount of heat that is carried away in the flue gasses is called stack loss.

9.4.1.4 Radiation Loss

Boilers vary in the amount of heat radiation. Inadequate insulation of the boiler and bad furnace design are the major causes of high radiation loss. Fire tube boilers usually have higher radiation loss than water tube boilers. Radiation loss normally increases when the load on the boiler decreases. A boiler with a radiation loss of 5 percent at 100 percent load could have a 20 percent radiation loss at 10 percent load. The installation of an oversized boiler to accommodate a possible future expansion can cause a continual excessive operating expense while the boiler has to operate at a fraction of full load capacity.

Boiler efficiency is normally from 70 to 80 percent for smaller boilers and somewhat higher for large boilers. It is most important that the heat transfer surfaces are free from deep soot of the fire side and free from scale on the water side. A 0.5 mm layer of scale can cause a drop of up to 11 percent in efficiency. The removal and prevention of scale is very important to maintain boiler efficiency and reduce steam cost. This is achieved by blowdown, chemical removal, and water treatment.

Soot and ash are removed daily with various mechanical and manual means including brushes and steam and air soot blowers.

9.4.2 Boiler Blowdown

When water boils, only the water vaporizes, leaving all the soluble solids in a more concentrated form in the remaining water. If it were possible to get a 100 percent condensate return, the concentration would remain the same since the return would be pure water. This is, however, not possible and makeup water is introduced with soluble solids in it. The boiler is designed to remove the vapor efficiently, thereby causing a rapid increase in the soluble solids concentration. At sufficiently high levels, the soluble solids will cause problems and the water in the boiler is partially drained. This process is called blowdown.

The purpose of blowdown is to decrease the amount of solids and sludge in the boiler water. It is important that the blowdown valves be free of pockets or dams that will allow sediment to collect. This is one of the reasons why globe valves are not used as blowdown valves. The blowdown system must have at least one of its valves as a slowly opening valve. The quickly opening valve should be located closest to the boiler while a more slowly opening valve can be closest to the steam box, quench tank, or blowdown tank. The more slowly opening valve will reduce the water hammer through the rapid vapor bubble collapse in the blowdown tank. The blowdown tank used with a boiler operates at atmospheric pressure and is vented. It is used to condense all the steam in the vapor/sludge mixture that will be ejected from the high-pressure boiler during blowdown.

Blowdown valves are normally positioned on the same side of the boiler as the water column gauge glass. It is important to remember that the blowdown piping should be as straight as possible. The number of times that blowdown takes place is dependent upon

1. The water treatment for makeup water
2. The amount of makeup water, which is dependent on the amount of condensate return
3. The quality of the makeup water
4. The type of boiler
5. The operating pressure of the boiler
6. The specific boiler application

The blowdown fitting on the boiler dictates the size of the piping on the blowdown system. The blowdown valves must be the same size as the piping. Blowdown can be done on a periodic basis such as once every shift or it can be triggered by measurements of the dissolved solids in the boiler water. It is wise never to exceed a soluble solid concentration of about 10,000 mg/L.

The scale formation can also be limited by chemical treatment of the boiler water. This process combined with blowdown should keep the formation of scale under control.

9.4.3 Water Treatment

Many dissolved substances that are found in water cause problems in boilers. Alkaline salts of calcium and magnesium can cause caustic embrittlement that can lead to metal failure. The water hardness caused by the salts of calcium and magnesium can be removed by hot or cold water precipitation using calcium hydroxide (hydrated lime) and sodium carbonate (soda ash), or sodium hydroxide (caustic soda) and sodium carbonate. It is also possible to remove the hardness by passing the water through base exchangers. The treatment is described in detail in Chapter 10.

9.5 Boiler Selection Criteria

In all countries steam boilers and other pressure vessels are regulated by a local agency. In America the American Society of Mechanical Engineers (ASME) formulates the regulations regarding the boiler industry. The ASME has codes of practice that specify boiler design, inspection, and quality assurance. Every boiler, just like all pressure vessels over a certain capacity, must carry an ASME stamp to ensure that it has passed inspection. In other parts of the world, the inspection is done by a government agency such as the department of manpower or labor.

On a more local level, each state has regulations that govern the use and operation of steam boilers. The regulations include the size, operating pressure, and the type of boiler. Another important aspect that should not be overlooked is the training and certification of the people that operate boilers.

9.5.1 Codes and Standards

A number of codes and standards, laws, and regulations that cover boilers and related pressure equipment must be considered when a system is designed. The regulations are dictated and enforced by a variety of organizations and agencies with an overall aim of making the workplace safe.

The company that insures the facility may dictate additional requirements to hedge its risk. To make this more acceptable to industry, companies offer lower premiums as incentive. The additional equipment is normally required to make the operation safer. Local and state authorities may require data on design features and control systems. Certain industries have additional criteria since the safety of the product can be dependent on the temperature time relationship of a process. Since the heat is provided by steam, the steam pressure and quality are very important. For the food industry, there are also regulations regarding the boiler room.

The blowdown system must be designed to ensure that the temperature of the effluent is such that it will not raise the temperature of the natural stream into which it drains. The temperature is normally measured a point downstream from the drain. Obviously, the sludge should not render the water unsuitable for sustained biological activity by changing the pH or poisoning of the water. The local authority could have temperature restrictions for water discharged into a sewer. In this case, a cooler is required after the blowdown tank.

9.5.2 Steam Pressure

Low-pressure boilers operate at pressures up to 103.4 kPa gauge pressure (GP) design and are typically used for heating applications. High-pressure boilers are typically used for process loads and can have an operating pressure of 517 to 4825.8 kPa GP. Most steam boiler systems require saturated steam.

Boilers are designed to operate at a specific pressure. The design pressure is higher than the operating pressure. The boiler is therefore overengineered with respect to the plate thickness. This is very important since the boilerplate will erode slowly with use. During inspection, the average thickness of the plate is checked.

To ensure safe operating pressures, the safety valves are set at or below design pressure. This is higher than the operating pressure for normal operation. If the pressure setting on the safety valve is close to the operating pressure, the safety valves will open regularly with a loss of efficiency.

Where superheated steam is required, steam heaters are installed. The increase in energy that is associated per kilogram of steam causes an increase in enthalpy. Since the steam is also drier, the steam quality is also enhanced. If very high pressures or superheated steam is required in an operation, it is best to install a water tube boiler.

9.5.3 System Load

The amount of steam that is required at a specific temperature and pressure is measured in kilojoules or kilograms. The boiler capacity must be designed to handle the maximum load of the system. The boiler turndown is designed from the minimum load requirement of the system. To determine optimum efficiency, the average load requirement is used for the calculations. The system load can be determined as heat load, process load, or a combination of the two.

9.5.3.1 Heating Load

To determine the maximum heating load of the boiler, one must consider the most extreme conditions. Once the heating load has been calculated for each area at the extreme conditions, the boiler size will be able to cope with normal loads. Remember to include all the uses of steam even if it is used for cooling where steam is used to run an absorption cooler. Losses of heat should also be considered, especially in long pipelines. Heating load normally utilizes low-pressure steam.

9.5.3.2 Process Load

Process load normally utilizes high-pressure steam. In this case the steam is used in manufacturing operations as a heat source. Process loads can be continuous, as in a hydrostatic retort where heat is supplied continuously and the demand fluctuations are moderate.

Batch processes are frequently used in industry and the steam demand fluctuates a lot. The steam demand is characterized by frequent short-time demands. The demand can be staggered but can also be simultaneous and that will cause a very large instantaneous demand. For batch operations, the boiler that is required will be much larger than what is dictated by average load. The boiler that is installed must be able to cope with large fluctuations in demand.

9.5.3.3 Combination Load

In many operations a combination of continuous and batch processes is serviced by one boiler. In this case all the considerations should be combined to make a decision on the size and type of boiler to install. In some practices where some continuous processes are not running 24 hours, the start-up for these processes is done when no batch demands will be operational. Heating up of a cold system takes a large amount of energy. If more than one system has to be brought into operation, the start-ups are staggered so as not to overload the boiler.

9.5.3.4 Load Variations

Since a factory normally has one main boiler and one standby boiler that are used when the main boiler is serviced, the boiler must be able to handle the variations in load. The selection of a boiler is dictated by load variation rather than by steam demand. Load can vary from day to day, by season, or instantaneously. It is also possible that it would be more economical to have more boilers in operation at certain times.

In many cases in the food industry, certain products are manufactured only at certain times of the year. The fruit processing industry is a good example of seasonal production with concomitant seasonal steam demand. The selection of the boiler will be dictated by the season with the highest load demand.

There is also a variation on a daily steam demand. When all or part of an operation runs for two shifts and the third shift is used for cleaning and sanitation of the system, it is obvious that there will be large daily fluctuations in steam demand. The highest daily and seasonal demand is used when determining boiler size, the number of boilers, and the turndown requirements.

Some peaks in demand occur because of intermittent use of certain lines or equipment that has a large heating requirement within the processing cycle. The equipment can be rated in steam demand per hour when it is running at a specified temperature. The demand for steam during start-up can be considerably higher than this rating. This can cause the steam system to be undersized.

9.5.3.5 Load Tracking

Load tracking is the ability of a boiler to supply a constant volume of steam at the required pressure during variable load operation. This ability depends on the type of boiler and especially on how quickly the burner and feed water system can react to changing demands.

The fire tube boiler has a large water capacity and is therefore very suitable to deal with instantaneous short-term large steam demands. When the variable load swings for considerable periods of time and when the pressure should remain constant, a water tube boiler is better suited to deal with it. The water tube boiler contains less water and can deal with variations much more rapidly.

9.5.4 Number of Boilers

In any situation, the number of boilers to feed the process must be considered. The process may run on one main boiler and, in some cases where zero downtime is considered, two identical boilers can be installed, although only one will be operating at any given time. Sometimes the second boiler is smaller than the main boiler since it is used only during maintenance on the main boiler. In cases where sudden product demand means that more than the normal processing lines are operating at the same time, the backup boiler can be brought into operation to increase steam capacity.

When noncritical heat systems fail, it will cause almost no loss in production and a backup is not warranted. The maintenance people will make the necessary repairs and bring the boiler into operation within a day. This setup is certainly no option in the northern states, where the winters are so severe that 18 hours without heat can cause considerable ice damage to all steam-heated systems and piping. When common sense dictates a backup boiler, laws and codes are normally in place to enforce it.

With continuous retort operations, a large amount of product is being processed during any specified time. If the steam should fail during the process, the loss caused by the failure can be considerable. To avoid this, more than one boiler should be in operation all the time with other boilers on standby. The standby boilers are also used during peak demand periods. The easiest way to make a decision regarding the number of boilers is to put a monetary value on the situation

9.5.4.1 Downtime Cost

If a fish processing plant has 25,000 cans of fish inside a hydrostatic retort and the product in the retort will be scrapped if steam is lost for more than 15 minutes, calculate the cost of the product and the cost to remove the waste and then compare that with the cost of an additional boiler.

In Minnesota we have to keep effluent ponds from freezing by heating them. With daytime temperatures of –20°C, it does not take very long for the pond to freeze when production stops because of a problem with the boiler. Imagine what happens to pipes full of condensate when they freeze up because the facility is no longer heated.

9.5.5 Boiler Turndown

Boilers have the ability to run at a fraction of their design capacity. Normally, the boiler will modulate down to enable it to run at 25 percent capacity. This means that a boiler that is rated at 400 kW with a 4:1 turndown burner will be able to turn down to 100 kW without turning off. This means that, at 25 percent load, the boiler will maintain constant operation. If the demand decreases more, the boiler will turn off and then turn on again once the pressure drops to the on cycle. This on and off cycling can happen every few minutes. With every cycle, the burner goes through a pre- and postpurge to remove heat from the boiler. This blows a large amount of heat into the stack and reduces efficiency dramatically.

It takes about 2 minutes to get a boiler back on line once it has turned off. If there is a sudden demand for steam during this off period, the burner will have to turn on before it can act on demand, so there may be a pressure loss in the system. A steam demand on the boiler when it has been turned down will result in an immediate increase in burner capacity with almost instantaneous response to demand.

It is most important that a boiler is bought with present demand as consideration. Buying an oversized boiler because of possible future demand will cause more problems than what it is worth. The most important considerations for the efficiency of a boiler are capacity and load requirements.

9.5.6 Fuels

Fuel cost is a large percentage of the overall costs in a food manufacturing plant. Fuel such as wood and coal are used, but in most industries stringent pollution control measures force the use of natural gas and light liquid fuel.

9.5.6.1 Fuel Oil

The advantages of fuel oil over other fuels are the following:

1. It is easy to adapt a boiler to use different fuel oils.
2. Handling requires less manual labor when compared to solid fuel.
3. There is more uniform heating.
4. The system requires less maintenance.
5. It eliminates dust and ash. (In Minnesota, ash is considered toxic waste!)
6. There are simple storage and handling systems.

The oil heating value varies from 38.2 MJ/dm³ for light oil to 42.7 MJ/dm³ for heavy oil (137,061 to 153,207 BTU/gal). Crude oil has an energy content of 38.5 MJ/dm³. High-viscosity oil is the most economical to use but it requires preheating to reduce the viscosity to ensure satisfactory atomization in the burner (see Table 9.1).

Fuel costs are variable and seasonal demand causes cyclic prices, so it is beneficial to have a boiler with a combination burner that can use liquid and gas fuel. This means that one can use the opportunity to utilize whatever fuel is cheapest at a

Table 9.1 Important Properties of Fuel Oils

Property	No. 2 (Light)	No. 6 (Heavy)
Flash point	38°C	60°C
Kinematic viscosity at 100°C	—	15 mm²/s
Sulfur content	0.50%	—
Relative density at 15°C	0.806–0.922	0.934–1.064
Boiling range °C	180–380	180–>500

specific time. It is also a great advantage when the gas or liquid fuel system requires maintenance to have an alternative fuel source. In many areas, there is also a peak usage period when fuel prices are higher per unit; switching to the alternate fuel source for the period will constitute large savings.

In some industries, part of the waste can be used as fuel in the boiler. This practice will reduce fuel cost and save on waste disposal. The waste streams are used in conjunction with standard fuels to ensure operating flexibility.

9.5.6.2 Gaseous Fuels

Gas is a clean and convenient fuel. Gas and oil can be used to fire automatic boilers and there is no problem with the disposal of ash. Food industries use natural gas (64.5 percent) as the preferred energy source, followed by coal (29.5 percent) and then oil (6 percent) (see Figure 9.3).

Natural gas is a perfect fuel source for heating purposes. It can be piped to wherever it is required and comes pressurized. The cost and rapid depletion of reserves are a downside. At present we deplete more than 1 percent of the known reserves per year. The energy content of natural gas is about 33.91 to 37.68 MJ/m^3. Natural gas must be supplied to the burner at a pressure of about 68.9 kPa.

The ignition of gas can take place before or after air has been mixed with it. The burners are thus classified as premixed or nonpremixed burners. When air is mixed with the gas prior to ignition, burning starts with hydroxylation. The aldehydes that form from the hydrocarbons are broken down when heat and more oxygen are added and H$_2$, CO, CO$_2$, and H$_2$O are formed. In this type of burning, there is no

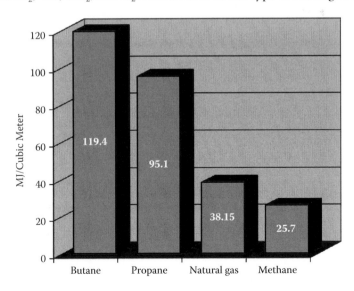

Figure 9.3   Energy values in megajoules per cubic meter for some gaseous fuels.

possibility to form soot. The flame temperatures are kept low to avoid the formation of nitrogen oxides.

The burner that is used most often is the partially premixed burner. The gas is mixed with some air so that it is flammable, but the mixture is still fuel rich. Secondary air is introduced around the flame holder to complete the oxidation so as to limit the formation of carbon monoxide. The lower temperature also limits the formation of nitrogen oxides.

9.5.6.3 Solid Fuels

In many industries all sorts of waste are burned to supply heat. Coal is, however, the solid fuel most often used in boilers. It is also the fossil fuel with the largest known reserves. There are three distinct forms of coal. Anthracite coal is black, with a shiny metallic luster. It has an energy content of 30.2 MJ/kg and burns with practically no flame at a temperature of about 2200°C. It has a bulk density of 800 – 930 kg/m^3.

Bituminous coal is black with a dull luster and is easily ignited. The energy content is similar to that of anthracite, but it is easier to ignite. There are three types of bituminous coal. Noncaking soft coal is preferred for industrial use. It will not fuse together in the furnace to form clinkers. The other types are caking and cannel coal. The bulk density of bituminous coal ranges from 670 – 910 kg/m^3.

Lignite has a brownish color. It has higher sulfur content than the other coal and an energy content of 23.2 MJ/kg. The bulk density is 640 – 860 kg/m^3.

9.5.6.3.1 Combustion of Solid Fuels

In suspension burners, pulverized coal is introduced into the furnace with a stream of air. The coal particle rapidly heats up and loses its volatile components. The volatiles will burn independently of the rest of the particle. The char particle can now burn and all that is left is the ash, which is composed of silicon, aluminum, iron, and calcium oxides. Suspension burning is normally used in large boilers with capacities greater than 45.4 Mg/h.

The general type of burner grate is the fuel-bed firing. The design that one frequently encounters is overfeed firing. Spreader stokers feed the coal onto a moving grate. Some of the coal will burn in the air while the rest will burn on the grate. About 40 percent of the coal is burned while it is in suspension. The residue is continuously removed from the grate.

9.5.6.4 Stack Emissions

The most important gaseous pollutants that are generated by burning are nitrogen oxides and sulfur oxides, plus particulate matter and unburned fuel. Nitrogen oxides (NO_X) refer to NO and NO_2 while sulfur oxides refer to SO_2 and SO_3.

9.5.6.4.1 Nitrogen Oxides

Nitrogen oxides are formed during combustion when oxygen combines with atmospheric nitrogen at high temperatures. A high oxygen concentration in the flame will cause a high concentration of nitrogen oxides. For this reason, the burning is staged such that the first part of the burning takes place in a fuel-rich primary zone. This oxygen-lean environment will limit the formation of nitrogen oxides. The heat is removed before the fuel mixture enters the second fuel-lean zone. Since the temperature is below 1810 K, the formation of nitrogen oxides will be limited even if sufficient oxygen is available. The staged burners can reduce nitrogen oxide emissions by as much as 30 – 60 percent.

Another approach is to recirculate the flue gas into the burner, thereby causing lower temperature burning that will reduce the formation of nitrogen oxides.

9.5.6.4.2 Sulfur Oxides

Sulfur in compounds is released during burning and the sulfur is converted to sulfur dioxide and sulfur trioxide. This conversion is independent of combustion conditions. The sulfur oxides will combine with water to form sulfurous and sulfuric acids. To prevent condensation of these compounds in the stack, a flue discharge temperature of about 420 K has to be maintained. This will cause limited heat recovery potential from the flue and loss of much energy in the stack. It will obviously still cause pollution.

The pollution can be limited if fuel with low sulfur content is used. Sulfur can be removed from flue gasses by reacting the gasses with limestone slurry.

9.5.6.4.3 Carbon Monoxide

During the oxidation of all hydrocarbons, carbon monoxide is formed before carbon dioxide is formed as the complete oxidation product. In optimized systems, the emission of carbon monoxide is very low since the oxidation reaction would have gone to completion. When the temperature is too low or when the amount of oxygen is limited, the emission of carbon monoxide can increase substantially. Optimizing the burner conditions is the best way to control carbon monoxide emission.

9.5.7 Replacement Boilers

If a boiler is selected to replace a boiler, the most important consideration is the floor space requirement. The footprint must fit and then the secondary space that surrounds the boiler should be considered. It is important not to forget the space requirement above the boiler. Some of the tubes in the boiler will be replaced during maintenance and this will have its own space requirement.

The operational characteristics of the existing boiler should also be kept in mind. If replacement is done to upgrade to a better system, it is important to consider all the selection criteria that were mentioned. Another important aspect is the stack position and location of the piping and utilities. Frequently, at this stage it becomes easier to replace a boiler with a newer model of the old one and the opportunity to upgrade to a more efficient system is lost.

Another consideration is access to the space where the boiler needs to be placed. Many boilers come in a kit form that is transported in pieces and assembled in place. Others are shipped as finished units that require a fair amount of access space into the building that will house the boiler.

9.6 Boiler Fittings and Accessories

Boiler fittings and accessories are mandated by ASME code to ensure that the operating environment is safe. As with any high-pressure processing vessel, the potential for catastrophic results exists if the boiler should fail.

9.6.1 Pressure Gauge

The pressure gauge will be attached to the boiler in a position where it will be directly connected to the steam space. A Bourdon tube type pressure gauge is normally used. The gauge is attached to the boiler with a loop so that there is a water bend between the Bourdon tube and the live steam. The Bourdon tube is a semicircular flattened tube that tends to straighten as the pressure increases. The change in the tube shape is reflected by an indicator hand.

9.6.2 Steam Pressure Regulating Valve

The steam in the boiler will be at a higher pressure than the steam required in the system. It is common for the design pressure of the boiler to be significantly higher than the design pressure for the rest of the system. It is therefore necessary to have a valve in place between the boiler and the system to reduce the pressure of the steam to the required pressure in the steam transfer system (Figure 9.4). The transfer system can operate at a higher pressure than the system that will be fed with the steam and so forth. Whenever a reduction of pressure must occur, a pressure-regulating valve is required.

If a pipeline strainer is installed, it is done before the pressure-regulating valve so that the water hammer effect on the valve can be limited.

9.6.3 Steam Trap

Some steam will condense somewhere in the pipe work and the steam will become contaminated with moisture. Steam traps (Figure 9.5) are devices that will remove

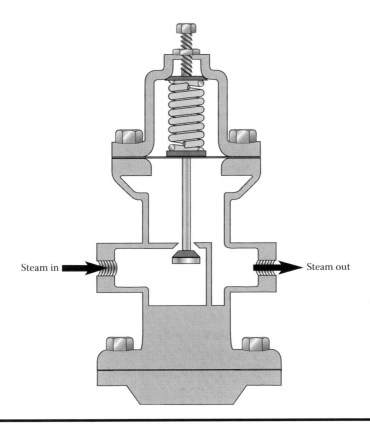

Figure 9.4   Pressure regulating valve.

the water from the steam line. Since a pressure-regulating valve will give erratic performance if it gets full of water, it is a good idea to install a steam trap before each pressure regulating valve.

9.6.4 Water Column Tube

The water column tube is a way that one can check the water level inside the boiler. There are many gadgets in place to ensure that the water level is maintained at the right level. These gadgets could fail but the water column is always right.

On high-pressure boilers, the water column gauge glass will normally have tri-cock valves to verify the water level inside the boiler. This means that the sight glass can be isolated for repairs or maintenance of seals.

The boiler will also have electronic sensing switches that will turn the boiler off in the case of too little or too much water. In both of these cases, the problem is frequently found in the boiler feed pump. The feedwater control system does not activate the feed pump when water is required or stops it when the boiler is full.

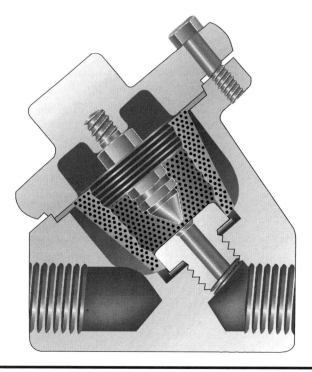

Figure 9.5 Steam trap.

Water column indicators are essential for reliable operation of a boiler. The maintenance of the water column by regular blowdown at least once a day is essential to keep the piping and the column clean of sludge.

Excessive accumulation of sludge in a boiler will prevent the proper operation of the low water cutout. The boiler will then function with insufficient water in it and that can result in damage to the pressure vessel.

9.6.5 Water Injection

Water is pumped into the boiler by means of a high-pressure makeup pump. It is important that the makeup tank that supplies the pump is maintained and full.

9.6.6 Safety Valve

The safety valve (Figure 9.6) is one of the most important accessories on a pressure system. The valve is designed to open when the set pressure is exceeded. It will remain open and blow until the pressure is reduced to 14 – 28 kPa below the pop-off pressure. All safety valves should be tested regularly to check that it is operational.

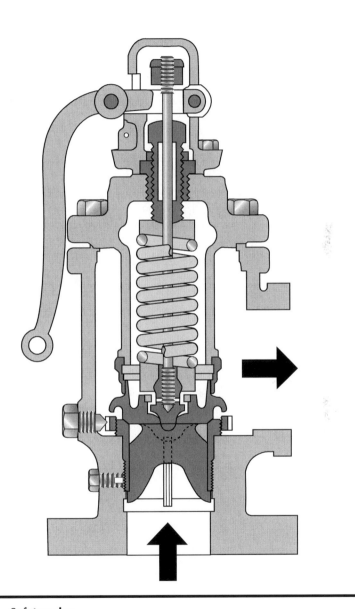

Figure 9.6   Safety valve.

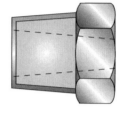

Figure 9.7   Fusible plug.

9.6.7 *Fusible Plug*

The fusible plug (Figure 9.7) is a little gadget of such simplicity that it is hard to believe just how effective it is. The plug is installed at a level in the boiler that is the lowest level that the water can drop to without damage to the boiler. The core of the plug is filled with a lead alloy that will not melt while it is covered with water.

When the water level drops below the level of the plug, there is no water to cool it down and the core will melt out of the center of the plug. When this happens, the steam inside the boiler is blown into the firebox. This will certainly get everyone's attention that there is some malfunction on the boiler. If at all possible, this should be avoided because it can be a time-consuming job to replace the plug (for example, the firebox needs to cool down).

9.7 Steam Purity

Whenever steam comes into contact with food, the steam should be culinary. It must not contain any contaminants that would be harmful if ingested. To fulfill the requirements for culinary steam, a secondary boiler should be set up. In this arrangement, the steam from the primary boiler is passed through a heat exchanger to heat the potable water in the secondary boiler.

Traps, filters, and pressure controls can be used on a normal steam line to deliver culinary steam. This can, however, only be done if no toxic chemicals are used to treat the boiler water. During the operation of a boiler, carryover is avoided as best as possible but it cannot always be prevented.

Foaming and priming can cause carryover. Contamination of the feedwater with suspended solids like oil or salts will cause foaming. Rapid changes in pressure will aggravate the problem. When rapid changes in pressure occur, cavitation and decavitation will cause small droplets of boiler water to contaminate the steam. This is called priming. Priming also occurs when the water level in the boiler is too high. In a boiler, the steam dome allows for proper separation of the steam and the water particles.

Chapter 10

Refrigeration and Freezing

10.1 Overview

The ancient Greeks and Romans used ice collected from the mountains to cool their drinks—obviously a luxury only affordable by a few very wealthy families. The ice was stored in deep pits lined with straw where the ice would keep well into summer.

The pioneer farmers in America built their milk-houses close to springs to use the cold water to cool the milk rapidly. The use of caves and cellars to store food is another well-known practice. Mechanical refrigeration was discovered in the mid-nineteenth century but was not commercialized until 1875. By the turn of the previous century, most households had an icebox—an insulated box that contained a block of ice to keep food cold. It was only after the First World War that a domestic refrigerator became the norm. After a very slow start, the application expanded to a way of life that we just cannot imagine doing without.

Modern refrigeration systems are capable of maintaining temperatures as low as –80°C. The low temperatures might be better to keep a frozen product for a long time, but maintaining these low temperatures is very costly.

Refrigeration and cooling are two terms meaning more or less the same thing. Refrigeration is usually used to describe the removal of heat from an enclosed space, while cooling is used for the removal of heat from a substance or object. Sometimes refrigeration is used to mean cooling to a temperature just above freezing point, while cooling means lowering the temperature somewhat. In both cases, there is a transfer of heat from a body or space to a medium that will then transfer the heat to another area or substance. The purpose of refrigeration is to lower the temperature of material to reduce the rate of chemical and enzymatic reactions. It will also reduce the natural metabolic rate of the tissue and bacteria associated with it to prolong the freshness of the materials.

In more affluent societies, consumers have the money to pay for fresh produce to be available at any time of the year. Products have to be kept refrigerated at appropriate temperatures to preserve freshness while they are shipped around the world. In Minnesota, fresh tropical fruit can be enjoyed at any time as long as the consumer is willing to pay for the cost of transport and refrigeration.

Freezing is a severe form of refrigeration that will cause the product to cool down to a point where the water starts to freeze. Freezing is used for long-term preservation of biologically active materials. At subzero temperatures, the metabolism of the food and that of any microorganisms are very low. Since the rate of chemical reactions is also greatly reduced, the products can be stored for extended periods without much deterioration.

Another application of refrigeration is to ensure comfortable working conditions with the use of air conditioning. The heat pump is a specialized refrigeration system that has the cold side inside the house in the summer; during the winter, the cycle is reversed with the cold side buried in the garden and the warm side in the house.

10.2 Natural Refrigeration

Stored ice was the principal means of refrigeration until the beginning of the twentieth century, and it is still used in some areas. Natural refrigeration is obtained when ice or ice-and-salt mixtures are used to absorb heat from the surrounding air. The heat of ice fusion is 334.9 kJ/kg at 0°C. In natural refrigeration every kilogram of ice that melts will absorb 334.9 kJ energy.

The specific heat of ice is 2.0934 kJ/kgK. This means that 2.0934 kJ of energy is required to raise the temperature of 1 kg of ice 1 K (1°C).

Example 10.1

How much heat is absorbed when 250 kg of ice at −15°C melts and the temperature of the resulting water increases to 5°C? The specific heat for water is 4.2 kJ/kgK.

Latent heat of ice:	250 kg × 334.9 kJ/kg = 83,725 kJ
Sensible heat from ice:	250 kg × 15 K × 2.0934 kJ/kgK = 7850.25 kJ
Sensible heat from ice water:	250 kg × 5K × 4.2 kJ/kgK = 5250 kJ
Total heat absorbed:	96,825.25 kJ

Example 10.2

When we expand the problem somewhat, the question can be asked as to how much water can be cooled down from 20°C to 5°C by the 250 kg ice.

$$X = \frac{96825.2 \text{ kJ}}{15 \text{ K} \times 4.2 \text{ kJ/kg K}} = 1536.9 \text{ kg}$$

10.2.1 Natural Refrigeration to Subzero Temperatures

When salt is added to ice, part of the ice will melt and the energy for the phase change has to come from the ice, thereby lowering the energy and thus the temperature of the ice. The temperature of the ice and salt mixture depends upon the amount of salt that is added. This was the way that homemade ice cream was frozen. In fact, it is still done today (see Table 10.1).

10.3 Solid CO_2

Sublimation cooling can be done with the use of solid carbon dioxide or dry ice. The phase change from solid to liquid takes place at a temperature of −78.5°C. This very low temperature will give very rapid heat flow because of the very large temperature difference as the driving force. Products can be cooled in crushed dry ice or the dry ice can be placed in acetone to give good thermal contact with products in containers.

Solid carbon dioxide freezing is frequently used for small-scale freezing to very low temperatures. It can also be used to freeze products to very low temperatures during transit.

10.4 Mechanical Refrigeration

Cooling caused by the rapid expansion of gases is the primary means of refrigeration today. The technique of evaporative cooling has been known for centuries, but the fundamental methods of mechanical refrigeration were only discovered in the middle of the nineteenth century. William Cullen at the University of Glasgow demonstrated the first known artificial refrigeration system in 1748. Cullen let ethyl ether boil into a partial vacuum; he did not, however, use the result to any practical purpose. In 1805 an American inventor, Oliver Evans, designed the first

Table 10.1 Temperature of Salt and Ice Mixtures

% Salt	Temperature °C	% Salt	Temperature °C
0	0	15	−11.7
5	−2.8	20	−16.9
10	−6.7	25	−23

refrigeration machine that used vapor instead of liquid. Evans never constructed his machine, but an American physician, John Gorrie, built one similar to it in 1844.

Ferdinand Carré developed a system using rapidly expanding ammonia in 1859. Ammonia liquefies at a much lower temperature than water and is thus able to absorb more heat. Carré's refrigerators were widely used, and compression–compression refrigeration became, and still is, the most widely used method of cooling.

In spite of the successful use of ammonia, that substance had a severe disadvantage: If it leaked, it was unpleasant as well as toxic. Chemists searched for compounds that could be used instead of ammonia and succeeded in the 1920s, when a number of synthetic refrigerants were developed. One of these compounds was patented under the brand name of Freon. It is a simple compound that is manufactured by substituting two chlorine and two fluorine atoms for the four hydrogen atoms in methane (CH_4). The resulting dichlorofluoromethane (CCl_2F_2) is odorless and is toxic only in extremely large doses.

A compression refrigeration system (Figure 10.1) consists of a compressor, a condenser, an expansion device, and an evaporator. The evaporated gas is compressed by a suitable mechanical device and then pushed through a condenser. Inside the condenser, heat is transferred to a fluid that absorbs the heat and moves away. The cooled high-pressure vapor can be passed through an expansion valve

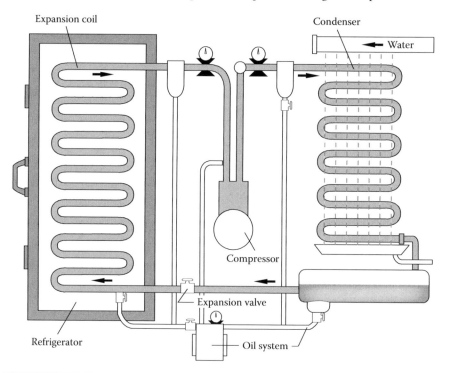

Figure 10.1 Compression refrigeration system.

to a region of lower pressure, where it will expand. The energy for expansion is removed from the area surrounding the evaporation coil.

The refrigerant in the reservoir underneath the condenser is under high pressure. It is forced to the region of low pressure through an expansion valve. The expansion valve opens automatically upon a temperature-sensed signal. It allows the refrigerant to pass into the evaporating coil that is at low pressure caused by the action of the compressor. The refrigerant absorbs heat from the surrounding area to enable it to evaporate. The area surrounding the evaporator tubes cools down.

The vapor then passes to a compressor. Here, the pressure is raised high enough so that the vapor will be able to condense once enough energy has been lost. The energy is transferred to the water or air that flows over the condensing system.

Both the compression and expansion of the vapor happen under isothermal conditions and follow the rule:

$$P_1 V_1 = P_2 V_2$$

The evaporation and condensation occur under isobaric conditions and follow the rule:

$$\frac{V_1}{T_1} = \frac{V_2}{T_2}$$

Many refinements are made to the basic operation to increase the efficiency of the system.

10.4.1 Capacity

The compressor in the system determines the capacity of the system. The refrigeration cycle depends on the amount of vapor that can be pumped per minute. There are many other factors that also influence the refrigeration capacity of the system. Like all systems, it is the balance of design that will ensure that each component part operates efficiently. It must be large enough to do what it should do but not oversized to the point of inefficiency.

10.4.2 The Evaporating Coil

The evaporating coil is that part of the system that absorbs heat. For the purpose of cooling, it is thus the operational part of the system such as the refrigerating coils of a cold storage room. In large systems where ammonia is used, the coils are made of iron or steel. In smaller units, the coils are frequently made of copper (when ammonia is not used). On compact coil designs, air is blown through the coils to promote air circulation and heat transfer. The "coil" can sometimes be long straight pipes, or a circular coil, or shell and tube construction.

The simplest construction is a plain expanding system. In this case, the expansion valve allows the required amount of liquid refrigerant to enter the coil pipe and the end of the pipe is connected to the suction side of the compressor. It is very important that the control of this system is such that liquid refrigerant will not be allowed to enter the compressor. If it does, it will cause inefficient operation and can even cause damage to the compressor.

10.4.2.1 Removal of Oil

Some oil is carried over from the compressor and collects in the cooling coils. It collects in the lowest part of the coil and should be drained off occasionally. Where a lot of oil is carried over, it will fill the lower coils, thereby reducing the refrigeration capacity. The compressor will work at a lower pressure, causing it to be inefficient. Problems with oil can be contained if oil is used that will not vaporize and that has a low pour point. Excessive oil in the system could be because of a broken or worn compressor ring. When the compressor runs against a vacuum for a long time, oil is atomized. Draining the oil from the evaporating coils at regular intervals will also assure improved performance.

10.4.3 Expansion Valve

The automatic expansion valve (Figure 10.2) controls the flow of refrigerant from the reservoir to the expansion coil. The high pressure in the system pushes the liquid refrigerant into the body of the valve. The refrigerant goes through a strainer and into a chamber where it presses against the piston and valve (3), forcing the valve to open. Refrigerant then flows through the valve opening (1) into the low-pressure side of the valve. A spacer washer (2) separates the arm (4) and the valve body (3). This action builds up pressure in a control chamber (5), forcing the diaphragm upward. The arm connection (4) allows the diaphragm action to move it and to close the valve (3). This ensures the balance of the system. The spring pushing on the diaphragm can be adjusted to balance the operation at any given pressure so that it will automatically throttle the main valve (3) and maintain the design pressure in the expansion coil of the system. This type of valve is not used with the flooded type evaporator.

The flooded type of evaporator is very efficient and trouble free. In this system, the coils are flooded. A float valve above the coils automatically maintains the desired level of refrigerant. The compressor suction is connected to the accumulator and not to the coil pipes. The accumulator acts as a trap to prevent liquid refrigerant from being passed to the compressor. It is a system that is simple and therefore easily automated. The accessories are a pressure regulating valve on the line to the compressor suction and an automatic stop valve, which positively shuts off the flow of liquid refrigerant to the float valve when the compressor is stopped.

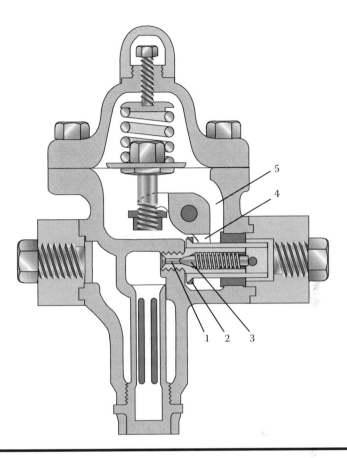

Figure 10.2 Automatic expansion valve.

10.4.4 Back-Pressure Regulator

By regulating the pressure in the coils, the temperature can be accurately maintained. The back-pressure regulator (Figure 10.3) regulates this pressure. The thermal efficiency of heat transfer from flooded coils is more than twice that of unflooded coils. It is of great benefit to regulate back-pressure when evaporation coils are in a brine tank, with fluid on the other side of the heat exchange surface, or in a continuous ice cream freezer, where the heat transfer must be at a very high rate.

10.4.5 Compressors

The compressor is the most important part of the system. In early years they were large, slowly moving piston machines that were very heavy. Modern compressors are built more along automotive lines of high speed and light construction.

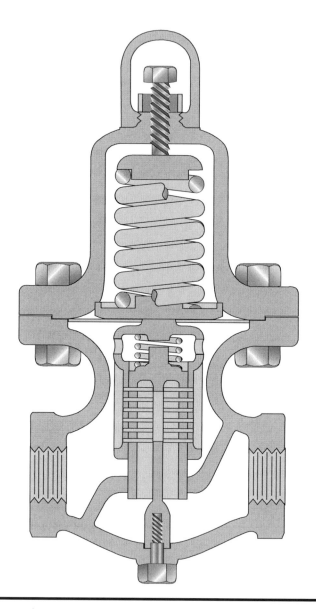

Figure 10.3 Back-pressure regulator.

Refrigerator compressors can be classified as positive displacement types or dynamic types. The positive displacement types are reciprocating and rotary compressors while the centrifugal compressor represents the dynamic type.

Positive displacement types of compressors are machines that will increase the pressure of a volume of vapor by reducing the volume. Reciprocating compressors (those with pistons) are normally selected for operations up to 300 kW refrigeration capacity. Centrifugal compressors are selected for operation greater than 500 kW. Screw compressors find application in between the ranges. The smaller vane compressors are frequently used for small-scale refrigeration such as in domestic systems.

Compressors for refrigeration must have large refrigeration capacity and power. Some reciprocating compressors have refrigeration capacity of 1 MW. The compressors are high speed, with up to sixteen single acting units on a direct-coupled drive system.

Centrifugal compressors, sometimes referred to as turbine compressors, serve refrigeration systems with capacities between 200 and 10,000 kW. The spinning impeller imparts energy to the gas that is compressed. Part of the kinetic energy is converted into pressure energy. In a multistage system, this conversion imparts high pressure.

To ensure proper running of a compressor, the pumping efficiency must be maintained and the bearings must be maintained and lubricated. The oil that is used must be able to remain liquid at even the lowest temperatures that can be encountered; yet, it should not break down at the high temperatures on the condenser side of the system.

10.4.6 The Condenser

The condenser is a heat exchanger where the heat that was absorbed in the evaporator is transferred to air or water. The combination of cooling and high pressure will cause the condensation of the refrigerant. Large systems frequently use double tube condensers that are cooled by water sprays. In small systems where low-pressure refrigerants are used, forced air circulation is used to remove the heat from the condenser pipes.

Where water-cooled systems are used in arid areas, the evaporative cooling from the water that evaporates from the hot tubes provides excellent condensation conditions. Here, very good cooling occurs with minimum water usage.

10.4.7 Automatic Control

Refrigeration systems rely on automatic controls for temperature, as well as condenser water and air flow, and flow of refrigerant. A float valve that is located in the liquid accumulator controls the flow of refrigerant in flooded systems. An electric stop valve will shut off the flow of refrigerant when the compressor shuts down. It will open as soon as the compressor starts up again.

The backpressure-regulating valve, which is in the suction line between the accumulator and the compressor, will maintain a certain pressure and thus the temperature in the evaporating coil.

The control systems on the condenser cooling water system will adjust the temperature of the condensed refrigerant. This will also control the pressure on the system. There should be an alarm system that will be activated when there is a lack of water for the condenser. A thermostat on the compressor can also be used to activate the pumps that circulate the condenser water.

An automatic expansion valve and thermostatic control can also be used to automate simple systems. They are frequently used on small cold rooms, counter refrigerators and freezers, and soft-serve ice cream freezers. The automatic expansion valve allows refrigerant to enter the evaporating coil or chamber when it is activated by a thermosensitive bulb that is attached to the coil or chamber where maximum refrigeration is obtained. An electric valve just before the expansion valve can be set to open when the compressor is running and to close when the compressor shuts down.

A very simple system that is used with low-pressure refrigeration systems utilizes a capillary tube to limit the flow of refrigerant into the expansion coil. This is used in place of the expansion valve. Combined with a small compressor, it works very well to control the flow of refrigerant.

10.4.8 Refrigerants

Refrigerants are any substance that acts as a cooling agent by absorbing heat from a body that needs to be cooled. Primary refrigerants describe the substances that are part of the refrigeration system that will absorb and release heat during the refrigeration cycle. Secondary refrigerants are cooled by the primary refrigerants and then distributed to cool down bodies remote from the refrigeration system. They are thus heat transfer fluids.

Good refrigerants must have the following properties:

1. Safe, nontoxic, and nonflammable
2. Environmentally safe
3. Compatible with the system and the fluids such as oil that can occur in the system
4. High latent heat
5. Low specific volume of vapor
6. Low viscosity
7. Low compression ratio
8. Fairly low pressures at operating temperatures
9. Low specific heat of liquid
10. High heat transfer characteristics
11. Stable under normal usage

Refrigerants are compared at standard conditions of –15°C for the evaporation temperature and 30°C for the condensing temperature.

10.4.8.1 Ammonia

Ammonia is a natural refrigerant that is still used in large-scale industrial refrigeration plants. Most fruits, juices, vegetables, and meat products will pass through at least one facility that uses an ammonia refrigeration system before they are purchased. It is used because of its excellent thermodynamic and thermophysical characteristics. Although ammonia has been used as refrigerant in large plants for more than 120 years, it is only now that its use is considered for smaller applications (Table 10.2).

Ammonia has the following desirable characteristics as a refrigerant:

1. Excellent thermodynamic and thermophysical properties
2. Higher energy efficiency in most temperature ranges
3. Known oil tolerance
4. Good tolerance to water contamination
5. Simple and immediate leak detection
6. Fairly low cost
7. Smaller pipe dimensions that can be used, thereby lowering plant investments

As with everything in life, ammonia has some disadvantages that should be considered:

1. Toxic at low concentrations in air
2. No tolerance to some materials like copper
3. Will not mix with most known oils
4. High discharge temperatures

An important consideration when ammonia plants are designed is the toxicity and flammability of ammonia. At high concentrations, ammonia is both flammable and toxic. It is flammable at concentrations of 15 to 30 percent by volume, a concentration that is unlikely in open air. At concentrations of 500 mg/L, it is toxic.

10.4.8.2 Other Refrigerants

The depletion of the ozone layer and the medical and environmental problems caused by it make Freon one of the environmentally most unwanted products. In fact, all substances containing chlorine and bromine seem to deplete the ozone layer. There are many new refrigerants that are used as a replacement for Freon. The

Table 10.2   Properties of Saturated Ammonia

Temperature (°C)	Saturation Pressure (kPa)	Enthalpy		Entropy		Specific Heat Vapor (kJ/kg K)	c_p/c_v
		Liquid (kJ/kg)	Vapor (kJ/kg)	Liquid (kJ/kg K)	Vapor (kJ/kg K)		
−70	10.89	−111.74	1357.0	−0.3143	6.9179	2.008	1.337
−60	21.85	−67.67	1375.0	−0.1025	6.6669	2.047	1.341
−50	40.81	−24.17	1392.2	0.0968	6.4444	2.102	1.346
−40	71.68	19.60	1408.4	0.2885	6.2455	2.175	1.352
−30	119.44	63.86	1423.6	0.4741	6.0664	2.268	1.360
−20	190.11	108.67	1437.6	0.6542	5.9041	2.379	1.370
−10	290.75	154.03	1450.4	0.8294	5.7559	2.510	1.383
0	429.4	200.00	1461.8	1.0	5.6196	2.660	1.400
10	615.0	246.62	1471.7	1.1666	5.4931	2.831	1.422
20	857.4	293.96	1479.8	1.3295	5.3746	3.027	1.451
30	1167.1	342.08	1485.9	1.4892	5.2623	3.252	1.489
40	1555.3	391.11	1489.8	1.6461	5.1546	3.516	1.538
50	2033.9	441.18	1491.1	1.8009	5.0497	3.832	1.602
60	2615.4	492.50	1489.3	1.9541	4.9460	4.221	1.687
70	3313.3	545.41	1483.9	2.1067	4.8416	4.716	1.801
80	4141.8	600.44	1474.2	2.2601	4.7342	5.374	1.960
90	5116.7	658.36	1459.0	2.4163	4.6209	6.302	2.192

major problem is to get them to work with the same efficiency in systems that were specifically designed for Freon.

The following compounds are part of the ICI list of possible Freon (R-12) alternatives

10.4.8.2.1 R-134a

R-134a, or 1,1,1,2-tetrafluoroethane, is widely accepted as the most appropriate alternative for R-12 in a wide application range for new and existing equipment. It is a good match to replace Freon in air conditioning systems operating at high and

Table 10.3   Properties of Saturated R-134a

Temperature (°C)	Saturation Pressure (kPa)	Enthalpy		Entropy		Specific Heat Vapor (kJ/kg K)	c_p/c_v
		Liquid (kJ/kg)	Vapor (kJ/kg)	Liquid (kJ/kg K)	Vapor (kJ/kg K)		
–100	0.56	75.71	337.00	0.4366	1.9456	0.592	1.161
–90	1.53	87.59	342.94	0.5032	1.8975	0.614	1.155
–80	3.69	99.65	349.03	0.5674	1.8585	0.637	1.151
–70	8.01	111.78	355.23	0.6286	1.8269	0.660	1.148
–60	15.94	123.96	361.51	0.6871	1.8016	0.685	1.146
–50	29.48	136.21	367.83	0.7432	1.7812	0.712	1.146
–40	51.22	148.57	374.16	0.7973	1.7649	0.740	1.148
–30	84.36	161.10	380.45	0.8498	1.7519	0.771	1.152
–20	132.68	173.82	386.66	0.9009	1.7417	0.805	1.157
–10	200.52	186.78	392.75	0.9509	1.7337	0.842	1.166
0	292.69	200.00	398.68	1.0000	1.7274	0.883	1.178
10	414.49	213.53	404.40	1.0483	1.7224	0.930	1.193
20	571.59	227.40	409.84	1.0960	1.7183	0.982	1.215
30	770.08	241.65	414.94	1.1432	1.7149	1.044	1.244
40	1023.89	256.35	419.58	1.1903	1.7115	1.120	1.285
50	1317.7	271.59	423.63	1.2373	1.7078	1.218	1.345
60	1681.5	287.49	426.86	1.2847	1.7031	1.354	1.438
70	2116.5	304.29	428.89	1.3332	1.6963	1.567	1.597
80	2633.1	322.41	429.02	1.3837	1.6855	1.967	1.917
90	3244.5	343.01	425.48	1.4392	1.6663	3.064	2.832

medium temperature ranges. For temperatures in the range of –20° to –30°C, a blend of R407 series such as 407D and R-500 should be used (Table 10.3).

10.4.8.2.2   R-407

The R-407-series refrigerants are blends of the three HFC refrigerants: R-32 (difluoromethane), R-125 (pentafluoroethane), and R-134a. R-407C was formulated as a good match to the existing HCFC refrigerant R-22 for use in air conditioning, chilling, and refrigeration applications and is the first of the R-22 alternatives to be used on a commercial scale. It is suitable for new and refurbished equipment usage.

R-407A is an HFC refrigerant that can be used in new equipment as an alternative to R-502 in supermarket refrigeration applications. It can replace R22 to some extent.

R-407B is also an R-502 alternative. It matches the thermophysical properties of R-502. It also has a reduced compressor discharge temperature that makes it a good choice for retrofitted applications where discharge temperature is of some concern.

10.4.8.2.3 R-32

R-32, or difluoromethane, is important as a blend for R-22 and R-502 alternatives. It is also a good industrial refrigerant that is almost nontoxic and noncorrosive. It has a high capacity similar to that of ammonia. Just like ammonia, it has a high compressor discharge temperature that requires appropriate handling technology.

10.4.8.2.4 R-125

R-125, or pentafluoroethane, is interesting as a blend component in both R-22 and R-502 alternatives. R-125 can also be used as a refrigerant, but its low critical temperature requires that the condensation section of the cycle be cooled with water (Table 10.4).

10.5 Thermodynamics of Vapor Compression

Change in enthalpy is driven by temperature, and pressure has very little effect on it. Entropy values in the refrigerant properties tables are used to check that it remains constant.

The enthalpy value is taken at the point (a) directly before the vapor enters the compressor. The vapor is compressed at constant entropy to the point where the compressed gas exits the compressor with enthalpy (b). The amount of heat that is lost in the compressor is relatively small and the assumption of an adiabatic process is reasonable. The other entropy values (c) are taken where the compressed vapor enters the condensation coils, the point where the condensed refrigerant leaves the coils (d), and the point where the refrigerant enters the expansion valve (e).

Since kinetic and potential energy values are insignificant in the vapor compression cycle, the first law of thermodynamics can be rewritten as

$$Q - W_s = \Delta H$$

Table 10.4 Properties of Some Refrigerants

	R-134a	R-407c	R-407A	R-407B	R-32	R-125
Saturated vapor density (kg/m^3)	5.26	5.67	3.94	5.57	2.98	6.76
Coefficient of volumetric thermal expansion (1/K 0–20°)	0.0028	0.0037	0.0034	0.0037	0.0036	0.0040
Trouton's constant (J/mol·K)	0.878	1.08	1.05	0.895	1.722	0.731
Latent heat of vaporization (kJ/kg)	177.33	193.9	178.8	145.3	270.22	112.6
Critical pressure (kPa)	4055				5816	3643
Critical temperature (°C)	101	86	83	76	78.35	66.3
Vapor/bubble pressure (kPa)	662	1190	1260	1330	1690	1378
Boiling/bubble point (°C)	−26.2	−44.0	−45.5	−47.3	−51.7	−48.5
Dew point (°C)		−36.8	−38.9	−42.9		
Composition (R32:R125:R134a)		23:25:52	20:40:40	10:70:20		

The quantity of heat transferred, Q, minus the work transferred, W_s, into a process is equal to the change in enthalpy, ΔH, of the product. For adiabatic compression Q = 0 and the work of compression is equal to the change in enthalpy:

$$-W_s = h_b - h_a$$

This is an equation for an ideal frictionless quantity and the equation must be divided by the compressor efficiency to get the actual work. For reciprocating compressors, the efficiency range is normally from 75 to 90 percent.

From the compressor to the condenser and through the condenser, no mechanical work is done and the heat change is equal to the change in enthalpy:

$$-Q_c = h_b - h_d$$

Q is negative because heat is lost from the system. The enthalpy change is an indication of change in sensible heat. From this point to the point where the liquid enters the expansion valve, the process has constant enthalpy and

$$h_d = h_e$$

The enthalpy changes in evaporation can be represented by the change from e to a. The change in heat is equal to enthalpy difference that reflects the refrigeration effect of the cycle:

$$Q_e = h_a - h_e$$

The useful refrigeration effect can be calculated by multiplying the enthalpy change with the mass floe of refrigerant (m_r):

$$Q_e = m_r(h_a - h_e)$$

where h_a is the enthalpy of the saturated vapor just before it enters the compressor (kilojoules per kilogram) and h_e is the enthalpy of the refrigerant entering the evaporator (kilojoules per kilogram).

The amount of energy that must be supplied by the compressor is represented by

$$W_a = m_r(h_b - h_a)$$

h_b is the enthalpy of the compressed gas that leaves the compressor (kilojoules per kilogram).

10.5.1 Coefficient of Performance

The coefficient of performance (CP) for cooling is an indication of the amount of useful refrigeration that can be obtained for an amount of energy used in the compressor:

$$CP = \frac{h_a - h_e}{h_b - h_a}$$

The coefficient of performance depends on the properties of the refrigerant and the condenser and evaporator temperatures. When the two temperatures move closer together, the work of compression decreases and the coefficient of performance increases. Most refrigerants will have CP values of 4.6 to 5.0 for condenser temperatures of 30°C and evaporator temperatures of –15°C.

When we assume that the gas is an ideal gas, the entropy of the saturated vapor at condenser pressure s_a is equal to the entropy of the compressed gas leaving the compressor s_b:

$$s_a = s_b$$

At constant pressure between the compressor and the condensation coil, s_c can be described as a function of the temperature change:

$$s_a = s_b = s_c + c_p \ln\left(\frac{T_b}{T_c}\right)$$

Solving for T_b will give

$$\ln \frac{T_b}{T_c} = \frac{s_a - s_c}{c_p}$$

$$T_b = (T_c) e^{\left(\frac{s_a - s_c}{c_p}\right)}$$

The enthalpy for b can be estimated by

$$h_b = h_c + c_p(T_b - T_c)$$

T_c is the absolute temperature of saturated vapor at condenser pressure, T_b is the absolute temperature of vapor leaving the compressor, and c_p is the specific heat of the vapor at constant pressure.

Example 10.3

Determine the difference in performance coefficient between ammonia and R-134a at standard testing temperatures. The evaporator is not superheated and the condenser is not supercooled.

At a temperature of –15°C the enthalpy and entropy of saturated vapor have to be calculated by averaging the readings at –20°C and –10°C. For ammonia, this is

$$h_a = \frac{1437.6 + 1450.4}{2} = 1444.0 \text{ kJ/kg}$$

$$s_a = \frac{5.7559 + 5.9041}{2} = 5.83 \text{ kJ/kg K}$$

At a condenser temperature of 30°C, we read from the table:

$$h_c = 1485.9 \text{ kJ/kg}$$

$$c_p = 3.252 \text{ kJ/kg K}$$

$$s_c = 5.2623 \text{ kJ/kg K}$$

$$h_d = 342.08 \text{ kJ/kg}$$

$$T_b = (T_c)^{\left(\frac{s_a - s_c}{c_p}\right)} = (303 \text{ K})e^{\left(\frac{5.83 - 5.2623 \text{ kJ/kg K}}{3.252 \text{ kJ/kg K}}\right)} = 303 \times e^{0.1745} = 360.77 \text{ K}$$

$$h_b = 1485.9 \text{ kJ/kg} + 3.252 \text{ kJ/kg K}(360.77 \text{ K} - 303 \text{ K}) = 1673.77 \text{ kJ/kg}$$

$$CP = \frac{1444 - 342.08 \text{ kJ/kg}}{1673.77 - 1444 \text{ kJ/kg}} = 4.80$$

Doing the same calculation for R-134a, we get

$$h_a = \frac{386.66 + 392.75}{2} = 389.71 \text{ kJ/kg}$$

$$s_a = \frac{1.7417 + 1.7337}{2} = 1.7377 \text{ kJ/kg K}$$

At a condenser temperature of 30°C we read from the table:

$$h_c = 414.94 \text{ kJ/kg}$$

$$c_p = 1.044 \text{ kJ/kg K}$$

$$s_c = 1.7149 \text{ kJ/kg K}$$

$$h_d = 241.65 \text{ kJ/kg}$$

$$T_b = (T_c)^{\left(\frac{s_a - s_c}{c_p}\right)} = (303\ \text{K}) e^{\left(\frac{1.7377 - 1.7149\ \text{kJ/kg K}}{1.044\ \text{kJ/kg K}}\right)} = 303 \times e^{0.0218} = 309.69\ \text{K}$$

$$h_b = 414.94\ \text{kJ/kg} + 1.044\ \text{kJ/kg K}(309.69\ \text{K}) = 421.92\ \text{kJ/kg}$$

$$CP = \frac{389.71 - 241.65\ \text{kJ/kg}}{421.92 - 389.71\ \text{kJ/kg}} = 4.60$$

The CP of ammonia is 4.8 compared to a CP of 4.6 for R134a. These values are somewhat higher than the values obtained from nonideal assumptions, but they are adequate as an estimate.

10.6 Secondary Refrigeration

Sweet water cooling is frequently used in the dairy industry. It is a term that means cooling with cold water as the cooling medium. It can only be used for applications that require cooling to temperatures above 0°C. On the other hand, it is a safe way of cooling substances that will suffer damage if the temperature is lower than 0°C. In the case of a leak in the system, there is only pure water added to the product so that it might be salvageable.

A direct expansion water cooler cools the water that is circulated to the material to be cooled. The system is efficient and has the safety factor that there is no possibility of freezing part of the product that is being cooled. The term "sweet water" describes the slight sweetness of cold water.

10.6.1 Congealing Tank System

When the sweet water system is modified to allow ice buildup around the coils, we get a congealing tank system. To allow for the ice buildup, it is necessary to open up the coils so that there is space for the ice buildup. The presence of ice increases the cold storage capacity of the system. Each kilogram of ice stores 334.9 kJ of refrigeration energy in its heat of fusion. This refrigeration capacity is useable as soon as the ice changes to water.

This system works well in operations where the cooling load varies and has many peaks in the load. Many processes that are of a batch nature have these peaks. When milk is delivered in bulk tanks, the milk has to be cooled to the desired temperature as soon as possible. The tankers will frequently arrive in clusters, taxing the refrigeration capacity at that time. For other parts of the day, there is no special demand on the system. The system can also be used if the refrigeration plant is too small to deal with peak loads that occur only once or twice a day for a relatively short time.

Congealing tank systems that use full-flooded direct expansion refrigeration systems are used for supplying cooling water at 0°C to 1°C. When colder temperatures are desired, the freezing point has to be lowered by the use of freezing point suppressants such as salt, alcohol, or glycerol.

10.6.2 Brine System

Brine systems are used in cooling applications below 0°C where it is undesirable or unsafe to circulate refrigerants such as ammonia or one of the hydrofluorides. Brine systems are not normally used in commercial refrigeration. They are used in industry for diverse applications where a solution with a high salt concentration is chilled and then circulated to do the required cooling at some point distance from the actual refrigerator system. The brines that are most frequently used contain sodium chloride and calcium chloride.

The main advantage of a brine system is that the refrigeration equipment is kept in an area that is remote from the place where the processing takes place (Figure 10.4). A leak in the brine system will cause less damage to the products than a leak of a primary refrigerant. The most important disadvantage is the higher energy cost to maintain the required temperatures at the point of application. The brine may also be corrosive to some metals.

Organizations like Greenpeace promote the use of brine systems with ammonia as primary refrigerant to replace the environmentally suspect hydrofluorides. Whether the hydrofluorides are safe or not does not matter; what matters is the public perception that industry does all it can to promote environmentally safe production.

The direct expansion of primary refrigerant in the expansion coils cools brine or sweet water. The cooled liquid is then circulated through a heat exchange system at the point where cooling is required.

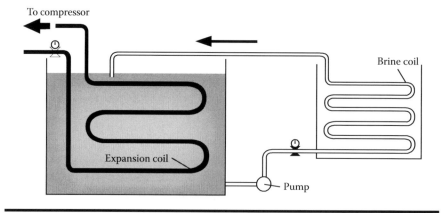

Figure 10.4   Indirect refrigeration with brine tank.

The extra heat transfer and the uptake of heat along the transfer line will cause higher energy requirements but reduce the danger of noxious or toxic refrigerants from contaminating a product or entering a workplace. It is also possible to cool large volumes of brine and thus store large amounts of refrigeration in a brine storage system.

The amount of refrigeration that can be stored in a brine tank can be calculated from the following formula:

$$Q_B = Mc\Delta T$$

Q is the amount of refrigeration stored that is equal to the product of the mass of the brine (M), the specific heat of the brine (c), and the difference in temperature of the brine.

Example 10.4

A tank holds 20,000 kg of brine containing 20 percent sodium chloride. The starting temperature of the brine is –7°C and the brine can be used for cooling until a temperature of –4°C is reached. How much useful refrigeration is stored in the tank?

$$Q_B = 20,000 \text{ kg} \times 0.829 \text{ J/kg K} \times (-4 - (-17)) \text{K}$$

$$Q_B = 215.54 \text{ kJ}$$

If the brine is stored in a tank with a height of 1.5 m and a diameter of 2.3 m, how much heat will be transferred from the air in the room and the brine if the room temperature is 20°C and the brine is at –17°C? Calculate the amount of heat transferred from the air to the brine:

$$h = c(\ T)^{0.25} = (1.3154 \text{ W/m}^2 \times 37 \text{ K})^{0.25} = 2.64 \text{ W/m}^2\text{K}$$

$$Q = Ah\ T = 2.3 \ \pi\text{m}^2 \times 2.64 \text{ W/m}^2 \text{ K} \times 37 \text{ K} = 705 \text{ W} \times 3600 = 2538 \text{ kW/h}$$

The brine must be able to operate at temperatures much lower than the temperature required for the operation. Remember it is still a heat transfer operation driven by the difference in temperature. For this reason, it is normally required that the brine temperature be at least 6°C colder than the required temperature of the process.

Table 10.5 shows the freezing point of NaCl brine. With CaCl2 brine, it is possible to go down to –48°C (–54°F) (Table 10.6).

Brine tends to be corrosive on equipment. To retard corrosion, the brine solution should be kept at pH 7. All air should be removed from the brine, and ammonia leaks into the brine should be prevented. Since the brine is full of electrolyte, dissimilar metals in contact with the brine will set up a battery with enhanced

Table 10.5 Properties of Sodium Chloride Brine

Specific Gravity at 4°C	Degrees Baumè at 15.6°C	Degrees Salo at 15.6°C	% Salt by Weight	Freezing Point (°C)	Specific Heat (J/kg K)
1.007	1	4	1	−0.11	0.992
1.015	2	8	2	−1.50	0.984
1.023	3	12	3	−2.33	0.976
1.030	4	16	4	−3.0	0.968
1.037	5	20	5	−3.78	0.960
1.045	6	24	6	−4.50	0.946
1.053	7	28	7	−5.28	0.932
1.061	8	32	8	−6.0	0.919
1.068	9	36	9	−6.72	0.905
1.076	10	40	10	−7.39	0.892
1.115	15	60	15	−11.0	0.855
1.155	20	80	20	−14.4	0.829
1.196	25	100	25	−17.5	0.783

corrosion of one of the metals. Even metals with different degrees of purity can cause this problem. Leakage of electrical current from electrical systems should be avoided. Chromates can be used to retard the corrosion, but any contamination of the chromate with the food product will make the food unusable.

10.7 Management

Leaks in a halide-based refrigerant system are located by means of a haloid lamp that operates on alcohol and is fitted with a rubber tube. If a halide compound with a concentration of 0.01 percent is introduced into the flame, the flame will turn a bright green color.

Keeping the high-pressure side pressure as low as possible and the low-pressure side pressure as high as possible optimizes the efficiency of a refrigeration system. To do this, the following important points should be considered.

To keep the high side pressure as low as possible:

1. Ensure sufficient condensation by efficient cooling
2. Keep the system clean and free from rust

Table 10.6   Properties of Calcium Chloride Brine

Specific Gravity at 4°C	Degrees Baumè at 15.6°C	Degrees Salo at 15.6°C	% Salt by Weight	Freezing Point (°C)	Specific Heat (J/kg K)
1.007	1	4	1	−0.5	0.99
1.015	2.1	8	2	−0.88	0.97
1.024	3.4	12	3	−1.39	0.96
1.032	4.5	16	4	−1.89	0.94
1.041	5.7	22	5	−2.39	0.93
1.049	6.8	26	6	−3.0	0.91
1.058	8	32	7	−3.61	0.90
1.067	9.1	36	8	−4.28	0.88
1.076	10.2	40	9	−5.11	0.87
1.085	11.4	44	10	−5.94	0.86
1.131	16.8	68	15	−11.0	0.795
1.179	22	88	20	−18.56	0.73
1.229	27	108	25	−29.89	0.685
1.283	32	128	30	−48.0	0.65

3. Keep the system free from noncondensible gases
4. Enlarge the condenser capacity if necessary

To keep the low side pressure as high as possible:

1. Have ample cooling surface
2. Keep the system free from noncondensible gases
3. Keep the system free from excessive oil accumulation in the cooling coils
4. Use fans to circulate the air over the cooling coils if possible
5. Use ample size pipe on all parts of the evaporation system

10.8 Storage Rooms

One of the main uses of refrigeration in many plants is the maintenance of a reduced temperature storage space. The space can be used to cool products down or to maintain the products at a reduced temperature. Frequently, the cooling down is done in one space where air movement will allow rapid removal of heat. Once the product is at the desired temperature, it is moved into another space with lower cooling capacity that will be able to maintain the low temperature.

In any ordinary space, the heat transfer taking place through walls, doors, floors, and ceilings is so large that the transfer of heat into the cold space will cause very high refrigeration costs. The refrigeration spaces are normally very well insulated to minimize heat transfer.

10.8.1 Insulation

The term "insulation" in refrigeration refers to materials that can be applied to walls, ceilings, floors, pipes, and tanks to reduce the transfer of heat. Still air is one of the better heat insulation materials. For that reason, most of the good insulating materials are lightweight and contains many small air cells (e.g., corkboard, foam, foam glass, urethane, and similar materials).

Convective and conductive heat transfer can only take place if molecules are present. A perfect vacuum will stop both forms of heat transfer, leaving only radiation heat transfer. The vacuum bottle with its silvered surface and its hollow body that is under vacuum is one of the best insulated vessels that we can manufacture.

When constructing a cold room, the materials used must be chosen to give the necessary strength and the required thermal insulation. Forklifts can cause substantial damage to a wall. The wall must therefore be constructed with materials that can withstand the impact at the level where it can occur (Table 10.7).

The proper insulation of cold storage rooms is of the utmost importance, since the cost of refrigeration as a method of preservation requires continuous input of energy. Sterilization with heat requires energy input once and the product is then shelf-stable. The room must be well insulated to maintain the required low temperature to protect the product in the event of an interruption of the power supply. The thickness of the insulation is calculated with the temperature difference between the ambient temperature and the cold room in mind. It also depends upon the type and cost of insulation.

It is important to design the cold room so that it is energy efficient. The upkeep of the structure should also be considered. When freezers were constructed of various organic materials and concrete, the system normally suffered structural damage if the walls were allowed to thaw and freeze. This was particularly critical with sharp freezers and rooms of −28°C to −40°C.

10.8.2 Prefabricated Polyurethane Panels for Cold Room Construction

Many prefabricated polyurethane panels are available. The panels are manufactured in lengths up to 12.2 m with a thickness of 25 to 300 mm. The facing of the

Table 10.7   Heat Conductivity and Density of Various Insulating and Building Materials

Material	Btu/h(ft)°F	W/m(K)	Density kg/m³
Air	0.014	0.024	1
Brick masonry	0.4	0.69	1760
Corkboard	0.025	0.043	160
Felt wool	0.03	0.052	100
Fiberglass	0.02	0.035	135
Glass	0.3–0.61	0.52–1.06	2500
Gypsum board	0.33	0.57	800
Polystyrene	0.019	0.032	38
Polystyrene	0.015	0.026	45
Polystyrene	0.023	0.040	25
Polyurethane	0.019	0.033	32
Mineral wool	0.025	0.043	145
Oak wood	0.12	0.21	700
Pine	0.087	0.15	500
Plywood	0.067	0.12	1100
Rubber	0.087	0.15	280
Steel, mild	26	45	7840
Steel, stainless	10	17	7950

panels is clad with polyester coated steel, PVC coated steel, stainless steel, stucco, aluminum, galvanized steel, etc. of various thickness. The density of the polyurethane is normally 40 kg/m³.

The cold room will have a locking system on the door. The door can be a normal hinged door, a double hinged door, a manual sliding door, an automatic sliding door, a hatch, a sectional roll-up door, or whatever else is needed. The door can also be equipped with dock seals that will allow loading of a trailer without loss of refrigeration.

When a cold room is constructed, the seals between any structural members are of great importance. The concrete curb is installed to prevent damage to the walls.

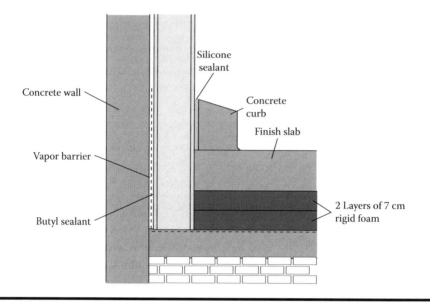

Figure 10.5 **Profile of the construction method of a cold room wall and floor.**

Since a large amount of heat can be lost through the floor, the floor is normally laid over insulation material. The cold side of the insulated wall must be made waterproof. If this is not well sealed, warm air-carrying moisture will seep in and condense. Wet insulation loses much of its insulating properties. Below freezing, the condensate will form ice that can destroy the insulation (Figure 10.5).

The ceiling of a cold storage room is insulated in much the same manner as the side walls. Floors, in addition to insulation, must have mechanical strength to carry the load of trucks, conveyors, and foot traffic. The insulation in the floors is frequently not as thick as in the walls and ceiling since the temperature of the soil is much lower than the temperature inside the room. The temperature difference is thus much smaller when compared to the ceiling or walls.

10.8.3 Cold Room Doors and Devices

Doors for cold storage rooms are made with heavy insulation and thus require heavy fittings. They are fitted with a double air seal so that warm air does not come in or cold air go out. Some companies specialize in door manufacturing.

Where there is a lot of traffic in and out of rooms, it is not advisable to have a door that requires constant opening and shutting. The door will be opened and left open until whatever needed to be done is done. This will cause a large loss of refrigeration. In these cases the opening is protected with plastic strip curtains that will allow the passage of forklifts but will reduce the loss of cold air to a degree. The see-through plastic strips will also allow the operators to see if anything will

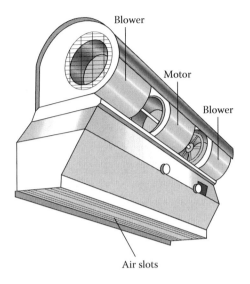

Figure 10.6   An air curtain.

obstruct the passage on the other side of the strip door. The strips will transmit 86 percent of light as compared to 92 percent for clear glass. The heavy-duty strips have the rigidity to hang straight when closed, but will give on impact. The strips overlap to give a moderate seal and some wind resistance.

Some companies also install air curtain doors (Figure 10.6) that allow free access of workmen and products, without severe loss of refrigeration. These systems are very useful in high-traffic areas. Air curtains have been developed for openings of any width up to 6 m in height. The air curtain depends upon a high velocity of air moving from the top to the bottom of the door to provide an air curtain, which will prevent the loss of most of the cold air through the door.

The air curtain has systems that will adjust the airflow in response to operating conditions such as wind and negative building pressure. It will also prevent the entrance of insects into the cold room. The air curtain is from 65 to 90 percent effective as a barrier to refrigeration outflow and 80 to 90 percent effective as an insect barrier.

10.8.4 Management of Cold Rooms

The management of cold rooms consists of maintaining a proper temperature in the cold room. Frequent inspection of heat transfer coils will ensure good heat transfer. The movement of material in and out of the cold room and efficient rotation to ensure first in, first out cycling is important

Many products, such as ice cream, require a stable storage temperature to prevent the growth of large ice crystals. For this reason, the product is frequently stacked in large blocks so that the sheer bulk will prevent temperature variation if there are short-time variations when products are moved into or out of the room.

10.8.4.1 Defrosting

During the regular operation of cold rooms, the coils become coated with a layer of ice, which decreases the heat transfer. Any moisture in the air inside the cold room will condense on the coldest body inside the cold room, which is the evaporation coil. The entrance of the cold room must be closed to prevent the entry of moist air and therefore the rapid icing up of the coils. Several methods are used to remove the ice from the coils.

The simplest and most inefficient way is to shut the refrigeration off and to open the door, letting in warm air. This is a slow process that causes considerable product deterioration. The ice can also be chipped away, which is a labor-intensive process in bad working conditions.

The compressor system can be reversed to move hot vapor through the coil, thereby thawing the ice directly around the coil so that it can be removed. This method is preferred since it requires little labor and will not warm the product appreciably.

Blower type cooling units have a special defrosting system, usually consisting of a hot gas line from the compressor combined with a flow of water or glycerol over the coils to remove the ice.

Chapter 11

Water and Waste Systems

Food industries utilize large quantities of water for every ton of raw material that they process. Some industries are within the municipal water reticulation system, so they will get treated water from a plant. Such municipal water supply treatment plants include facilities for storage, treatment, transmission, and pressurized distribution. The specific facility will be designed to treat the particular water. The design is also influenced by the quality of the water, the particular needs of the user or consumer, and the quantities of water that must be processed.

11.1 Water Quality

Water is used for many different purposes in food processing plants. Many of the processes that utilize most of the water are at the input side of the process, where incoming materials require cleaning. Water is used for washing products, fluming, cooling, condensing, cleaning, sanitizing, heating, steam generation, and cleaning. Water is also an important added ingredient in many processed food products. In some products, like beer, pop, and bottled water, it is the main ingredient. Food industries are frequently located where they are because of the supply of water. Beer is one of the products that are very dependent upon water quality for final product taste and quality. An adequate supply of safe and high-quality water is essential for the food industry.

The water that is processed by municipal treatment systems has been proven safe over a long period of time. Most of us take water safety for granted. Many food plants operate their own wells and treatment systems and have to ensure their own water quality and quantity. Products frequently dictate more than just safe water and will require additional treatment to ensure product suitability. The pop industry is one of these where water needs to be treated to remove many ions that will interfere with the taste of the product. Such special water treating systems are commonly found in food plants to obtain water with the characteristics desired for special purposes. Water treatment as feedwater for boilers is one such special treatment.

11.2 Potable Water

When any precipitation falls, it has to go through a rather impure air layer where it picks up all sorts of substances and organisms. Oxygen, carbon dioxide, and sulfur dioxide are gases that will readily dissolve in water. Dust particles and airborne organisms will be suspended in the water. Once it drops to earth and runs through our less than pristine environment, it will dissolve minerals and man-made substances like insecticides, herbicides, etc. More solids and clays will become suspended until we have a solution with a fairly stable colloid flowing through the countryside. In lakes, bogs, and swamps, water may sustain a large algal population, causing large amounts of gums to be released in the water. This will cause natural foaming and it is not uncommon to see large patches of stable foam on the Red River in early spring. The water will also gain color, taste, and odor from decaying vegetation.

The water that manages to trickle through the crust to become groundwater usually acquires more dissolved minerals because of its long contact with soluble minerals. As it moves through the different subterranean formations, it can absorb gases such as hydrogen sulfide and methane. Human activities and the effects of pollution influence the quality of surface and groundwater.

11.2.1 Water Safety

Many dissolved and suspended substances in water can influence the specific functional usability of water. The main consideration regarding water is its safety for human consumption. All public water supplies are under the jurisdiction of the federal Environmental Protection Agency (EPA). In many states the local state agencies will serve as the enforcement authority. The EPA has many regulations that specify the adequate monitoring of water supplies, the maximum contaminant levels, testing methods, enforcement of regulation, and reporting procedures for noncompliance.

Maximum dissolved contaminant levels for inorganic substances allowed in drinking water are shown in Table 11.1.

The amount of fluoride in water is also monitored and the amount of permissible fluoride is dependent upon the average daily temperature (Table 11.2).

Table 11.1   Maximum Inorganic Chemical Contaminants

Contaminant	Level (mg/L)	Contaminant	Level (mg/L)
Arsenic	0.05	Barium	1.00
Cadmium	0.010	Chromium	0.05
Lead	0.05	Mercury	0.
Nitrate (as N)	10.00	Selenium	0.01
Silver	0.05		

Table 11.2   Maximum Fluoride Levels in Water

Temperature (°C)	Fluoride Level (mg/L)
12.0 and below	2.4
12.1–14.6	2.2
14.7–17.6	2.0
17.7–21.4	1.8
21.5–26.2	1.6
26.3–32.5	1.4

11.2.1.1 Health Concerns

The five impurities in water that cause public health concerns are organic and inorganic chemicals, turbidity, microorganisms, and radioactive materials. Organic contaminants include insecticides, herbicides, fungicides, industrial solvents, soaps, and organic solvents. Inorganic contaminants include arsenic, nitrate, fluoride, and toxic metals such as chromium, lead, and mercury. All these substances can be harmful if they are present above certain concentrations in drinking water. Fluoride is harmful in large quantities but can be added as trace amounts to water supplies to help prevent tooth decay.

11.2.1.1.1 Turbidity

Turbidity refers to the permanent colloid caused by very small particles of silt, clay, and other substances suspended in water. For drinking water even slight turbidity is objectionable. The turbidity will also interfere with the action of chlorination as our principal method of disinfecting the water. It can react with most of the available chlorine or it can keep the pathogenic organisms from being killed by the chlorine. Groundwater is normally less turbid because of the natural filtration process to become groundwater. Surface waters are frequently high in turbidity, especially when rivers are called red, yellow, and whatever where the color is caused by colloidal clay particles.

11.2.1.1.2 Microbiological Concerns

Natural water is very rarely sterile. It may be safe to drink because the organisms in the water do not cause any disease. For practical purposes and to err on the side of safety, all water used for drinking purposes or as ingredients will be treated to assure safety. Water-borne diseases are not common in developed countries because of careful monitoring of supplies, good sanitation practices, and adequate chemical treatment. The pathogenic bacteria listed in Table 11.3 can live in water.

When water is contaminated with fecal matter, bacteria of the coliform group serve as an indicator organism. *Escherichia coli, Aerobacter aerogenes,* and *Streptococcus pyogenes* are found in fecal matter of warm-blooded animals. Their presence in water indicates a strong possibility of sewage pollution. The organisms that were mentioned will cause only a mild or no disease; they are indicators of the possibility of salmonella and shigella, which that can cause much more serious disease. Experience has shown that treatment that will kill the coliforms will kill the pathogens and render the water safe.

Water contaminated with human waste always contains coliforms, and it is likely to contain pathogens excreted by infected individuals in the community. One pathogen that is of special concern is the virus that causes hepatitis. It will survive* in water and it will even survive in 100°C water for 5 minutes. If coliforms are not found in the water, we assume that the water is also free of pathogens.

Table 11.3　**Pathogenic Bacteria Found in Water**

Organism	Disease
Salmonella typhosa	Typhoid fever
Shigella dysenteriae	Bacillary dysentery
Entamoeba histolytica	Amoebic dysentery
Salmonella paratyphi	Paratyphoid A
Salmonella schottmuelleri	Paratyphoid B
Vibrio cholerae	Asiatic cholera

---

* Because viruses cannot multiply on their own, they do not fit the definition of being alive. They are active to infect other living cells or they are not active. This is a problem of the definition that defines "alive." Can we kill something that is not alive?

When the coliform count is low, there is little change that pathogens are present in the water.

Various testing methods are used to indicate and enumerate the presence of coliforms in water. The number of organism must be less than 1 in 100 mL. This number is only achievable in treated water. Water must be checked on a daily basis for coliform and chlorine residual. A free chlorine residual of at least 0.2 mg/L must be maintained throughout a water distribution system. This will not always inactivate viruses.

If considerable sewage contamination occurs in a water source, the source should be eliminated if possible. If it is not possible, the contamination should be prevented in the future. The water source should then be tested more regularly so that steps can be taken to avoid product contamination.

11.2.1.1.3 Radioactive Materials

Radioactive materials are seldom found in water supplies since the use and disposal of radioactive materials are very well monitored. Possible sources of pollution are mining waste, nuclear power plant accidents, and research institutions where someone washed radioactive materials down the drain.

11.2.1.2 Aesthetic Concerns

The recent proliferation of various water-filtering systems used in homes is a good indicator that consumers want good drinking water that is free from color, odor, chemicals, and contaminants. The listed properties pose no health risk and can be seen as aesthetic considerations. Decaying organic materials can cause a brownish yellow hue in water. Natural dissolved organic substances or gases may cause taste and odor. When water contains dissolved metals, they can cause a strange metallic taste and may stain laundry and plumbing fixtures. Excessive chlorides give the water an objectionable salty taste. Excessive minerals cause a lime taste or a salt taste.

11.3 Substances in Water

Water is very seldom pure. Trying to get purified water normally means distillation of water in more than one step. The final distillation occurs in a special still. Water is a good solvent and it is therefore full of solubles.

11.3.1 Hardness

Hardness in water is expressed in terms of equivalent calcium carbonate measured in milligrams per liter (ppm or mg dm^{-3}). Hardness is not harmful to humans but

Table 11.4 Hardness Scale

Hardness Description	Calcium (mg/L)
Soft	0–60
Moderately hard	61–120
Hard	121–180
Very hard	Over 180

it is rather objectionable if calcium deposits build up on everything. Many consumers in America require water-softening units to utilize the hard water in their geographic location. The water hardness in the United States varies from 17 to 6000 mg/L calcium. The average range is from 51 to 850 mg/L (Table 11.4).

Hardness is the presence of dissolved minerals in water and especially salts of calcium and magnesium. The most common dissolved minerals are calcium bicarbonate, magnesium bicarbonate, calcium chloride, magnesium chloride, and magnesium sulfate. Bicarbonates of calcium and magnesium are soluble and we refer to their presence as temporary hardness. When water with bicarbonates is heated, the bicarbonates will change to carbonates. The carbonates are not very soluble and will deposit out of solution, forming scale or precipitate. Scale on heat transfer surfaces decreases heat transfer. When scale accumulates in pipelines, it will eventually reduce the pipe diameter sufficiently to reduce the flow of water.

Bicarbonates are formed when carbonic acid dissolves limestone. Carbonic acid is formed when carbon dioxide dissolves rainwater as it filters through decaying vegetation:

$$CO_2 + H_2O \leftrightarrow H_2CO_3$$

$$CaCO_3 + H_2CO_3 \leftrightarrow Ca(HCO_3)_2$$

When the water is boiled, this reaction is reversed and the low solubility calcium carbonate will precipitate. At the same time, reaction with sulfates and chlorides in the water will take place to cause permanent hardness. This will cause a hard scale in pipelines or on equipment.

11.3.2 Iron

After magnesium and calcium bicarbonates, iron is third in line for causing problems in water supplies. Iron is frequently found in acidic water with manganese and sulfur. This combination makes treatment more difficult. In most cases the iron content of water is less than 5 mg/L. In severe cases it can be as high as 60 mg/L. Very low iron concentrations can still cause problems. In concentrations of 0.3 mg/L, it can still cause stains on equipment, walls, and floors. For final rinse water and sanitizing solutions, iron content in excess of 0.3 mg/L will cause problems.

In well water the iron will be in the ferrous form. This will be clear but after contact with air or chlorine, it will oxidize to ferric compounds that are reddish brown. In surface water, the iron is normally in the ferric state. Both the ferrous and ferric states of iron are objectionable. Iron-rich water will provide a growth medium for certain bacteria that require high iron concentrations. The precipitate formed by the growth of these organisms can cause blocked pipes.

11.3.3 Manganese

Manganese can be found in well water with iron. In surface water manganese dioxide is formed and precipitates. Manganese is present in small amounts, normally not exceeding 3 mg/L. Even at that low level, it will cause black stains that are difficult to remove.

11.3.4 Nitrates

Nitrates in water are normally caused by water running over naturally nitrate-rich soil. The nitrate concentration normally does not exceed 4 mg/L. Higher concentrations of nitrate usually indicate some problem. One of the problems is that nitrogen-rich substrates will cause increased levels of nitrate in the water. Frequently, the origin of the nitrogen-rich substrate is sewage pollution of the water source.

11.3.5 Chlorides and Sulfates

High concentrations of chlorides in water will impart a brackish taste to the water. In some areas, salt deposits cause most of the well water to have elevated chloride content. At low concentrations, chlorides will improve the taste of water. At high concentrations of about 50 mg/L, the water will have an objectionable taste. Water with high concentrations of chloride is also more corrosive.

Sulfates in water produce scale that is very difficult to remove. Calcium sulfate or gypsum also imparts a bad taste to water. Magnesium sulfate has some laxative properties and this may prove a discomfort to people not accustomed to the magnesium sulfate in local water.

11.3.6 Gases

It is normal to find oxygen and carbon dioxide in surface water. Without it, no aquatic life would have been possible. Dissolved oxygen promotes oxidation of metals, especially iron, brass, and galvanized metal. Carbon dioxide dissolves in water to form carbonic acid. This will increase the corrosiveness of the water.

Hydrogen sulfide is noticeable at a concentration of less than 1 mg/L. It causes disruption of the exchange action in exchange resins and will promote tarnishing of certain metals. In food plants, the hydrogen sulfide has to be removed.

11.3.7 Silica

Water flowing over granite rocks and sediments will dissolve small amounts of silica. The silica will not cause any problems except in high-pressure boilers, where it can form a very hard scale. Treatment of feedwater is required to prevent the formation of scale. In boilers operating at pressures below 1030 kPa, scale formation can be prevented by adequate blowdown of the boiler.

11.4 Treatment of Water Supplies

Water from municipal plants may be safe for drinking but will not satisfy the specifications for industrial water. Most industries will have to treat all or part of the water for specific usage. Feedwater for boilers is of particular importance and requires many treatments. Water used as ingredients obviously requires treatment specific to the product. This normally includes removal of odors, color, chlorine, and turbidity. Water used in cooling towers or can cooling systems requires treatment to inhibit corrosion of surfaces and deposits or stains on surfaces. Treatment will include removal of hardness and specific ions such as iron and manganese, removal of gases, and addition of substances such as chromate.

Each operation will have a unique water treatment program that is set up for the product, the water quality, and the equipment that is used. A procedure that works well for one product or system may be totally unsuited for another system or product. There are many specialists that will be able to help to design a system for a specific requirement.

11.4.1 Treatment to Remove Turbidity

Turbidity is caused by suspended solid particles in water. The particle size will range from 1 to 200 μm (micrometer) for colloidal suspensions and about 100 μm for fine sand and silt. If water is held undisturbed for long periods, particles with diameters larger than 10 μm tend to settle out. To produce clear water, the particles in colloidal suspension need to be removed. Colloids are normally very stable* and treatment with a coagulant is needed to aggregate the particles. Once an aggregate of sufficient density and size has formed, it will settle out. Coagulants release positive ions when they dissolve in water. The ions attract the negatively charged

---

* The casein in milk that produces its white color is a good example of a stable colloid.

colloidal particles to form a jelly-like floc. The floc assists in aggregating more colloidal particles. Rapid settling properties increase the clarifying efficiency. Adding filter aids that will speed up floc formation and settling can sometimes improve clarifying.

Chemicals that are commonly used as coagulants include ferric sulfate ($Fe_2(SO_4)_3$), ferrous sulfate ($FeSO_4$), alum ($Al_2(SO_4)_3$), and sodium aluminate ($Na_2Al_2SO_4$). The coagulants react with alkalis in the water to form hydroxides ($Fe(OH)_3$ or $Al(OH)_3$); this forms the floc. Reactions for the coagulants with sodium bicarbonate in the water are as follows:

$$Fe_2(SO_4)_3 + 6\ NaHCO_3 \rightarrow 3\ Na_2SO_4 + 2\ Fe(OH)_3 + 6\ CO_2$$

$$Al_2(SO_4)_3 + 6\ NaHCO_3 \rightarrow 3\ Na_2SO_4 + 2\ Al(OH)_3 + 6\ CO_2$$

Sodium aluminate is a fast acting coagulant and can be used in much lower concentrations than filter alum and is thus more economical. The reaction of sodium aluminate with water is as follows:

$$Na_2Al_2O_4 + 4\ H_2O \rightarrow 2\ Al(OH)_3 + 2\ NaOH$$

The more popular practice is to use organic polymers to facilitate water clarification by coagulation. Cationic polymers carry a positive charge that neutralizes the negatively charged particles. The nonionic part of the polymer then aids in floc formation. These materials are a group of polyamines and are frequently referred to as polyelectrolytes. Activated silica and special clays may be used to enhance floc formation and settling of particles.

11.4.2 Softening Hard Water

When water is treated to remove turbidity, some of the hardness is also removed. Sodium aluminate will not only act as a coagulant, but will also add alkalinity to the water to reduce some of the noncarbonate hardness in the water:

$$Na_2Al_2O_4 + 4\ H_2O \rightarrow 2\ Al(OH)_3 \downarrow + 2\ NaOH$$

$$MgSO_4 + 2\ NaOH \rightarrow Na_2SO_4 + Mg(OH)_2 \downarrow$$

When sodium aluminate is used as a coagulant in turbidity treatment, the amount of lime that is required to reach the softening pH in the cold lime method of softening can be reduced by about 10 percent.

11.4.2.1 Cold Lime Method

Cold lime softening has been used over the past 150 years. In the treatment process, calcium oxide (CaO) is added to the hard water to form calcium hydroxide that reacts with magnesium and calcium bicarbonates and free CO_2 to form insoluble calcium carbonate and magnesium hydroxide.

$$CaO + H_2O \rightarrow Ca(OH)_2$$

$$CO_2 + Ca(OH)_2 \rightarrow CaCO_3 \downarrow + H_2O$$

$$Ca(HCO_3)_2 + Ca(OH)_2 \rightarrow 2\ CaCO_3 \downarrow + 2\ H_2O$$

$$Mg(HCO_3)_2 + Ca(OH)_2 \rightarrow Mg(OH)_2 \downarrow + 2\ CaCO_3 \downarrow + 2\ H_2O$$

Magnesium hydroxide has good flocculating properties and helps to precipitating the calcium carbonate particles. Any excess lime is converted to calcium carbonate by addition of carbon dioxide to the water as it leaves the primary settling tanks. At the same time, the magnesium hydroxide will be converted to magnesium carbonate. After this carbonation step, sodium carbonate is added to react with noncarbonate salts to form calcium carbonate that will precipitate. The other compounds are soluble and will stay in the water:

$$Na_2CO_3 + CaSO_4 \rightarrow Na_2SO_4 + CaCO_3 \downarrow$$

$$Na_2CO_3 + CaCl_2 \rightarrow 2\ NaCl + CaCO_3 \downarrow$$

After this treatment, the water hardness will be about 80 mg/L, placing it in the moderately hard category. Sand and gravel filters are used to remove the precipitated salts from the water.

11.4.2.2 Ion Exchange

Many ion exchange processes are used to treat water. Water softening is one of the common treatments. In the system, calcium and magnesium ions in the water are exchanged with sodium ions while the water flows through a resin bed. After a time, the sodium ions will become depleted and the resin requires regeneration, where the calcium and magnesium ions are removed and sodium ions are replaced. This is done by pumping brine solution through the resin. This process will not reduce turbidity or remove other chemicals.

There is a more sophisticated exchange deionization process available for water treatment. In this case the calcium and magnesium ions are removed and they are replaced with water, leaving no undesirable increase of sodium concentration.

Table 11.5   Ion Exchange Resins

Product	Applications
Amberjet, Amberlite, and Duolite ion exchange resins	Water softening, dealkinization, deionization, silica removal, desalination, condensate polishing, sugar purification, metal recovery, waste and water treatment
Amberlite polymeric adsorbents	Food and beverage processing
Amberlyst ion exchange resins	Removal of impurities from chemical intermediates, including mercaptans, acids, and phenolics; catalysis; use in manufacture of alkylated phenols and gasoline octane improvers
Ambersep ion exchange resins	Processes for making ultrapure water for use in power plants or the manufacture of microelectronic components

The resin is normally augmented with an activated carbon filter that will remove chlorine, lead, boron, and all sorts of other dissolved and suspended materials (Table 11.5)—a wonderful system for drinking water but somewhat over the top for industrial applications.

11.4.2.2.1 Base Exchange

The base exchange resins used in the food industry utilize ion exchange beds of natural or synthetic zeolites that are hydrous silicate or styrene-based resins. In sodium exchange resins, sodium from the resin is exchanged for calcium and magnesium in the water. Some exchange of iron, copper, manganese, and aluminum will also occur.

Many modern resins are based on a sulfonated styrene divinylbenzene structure. There are resins for almost every kind of application and the selection is based on the water analysis, the operating temperature, and the required outflow water quality. Resins are very costly and durability under usage conditions is one of the important selection criteria.

The sulfonic groups in styrene-based resins are responsible for the cation exchange property. The resins are able to operate over a wide pH range. The water flows through a bed of spherical resin beads. After some use, the resin is regenerated in three steps:

1. The resin is backwashed to remove solids that could have accumulated on the resin bed surface. The beads are suspended in the upward flow that will also loosen any packed bead clumps.

2. A saturated salt solution is introduced on top of the resin and allowed to flow downward. The salt will remove calcium and magnesium ions and replace them with sodium ions. This leaves the resin in the sodium condition.
3. In the final step, the resin is rinsed to remove excess salt, and calcium and magnesium chlorides. The softener is ready to return to service.

In plants, there is normally more than one softener so that some will be regenerated while others are operational. Regeneration can take from 35 to 70 minutes. In most plants, the regeneration of softeners is completely automated. This ensures a continuous supply of soft water at the lowest operating cost.

11.4.3 Water Demineralizing

Water softening is a process that is very common in the food industry. In some sectors, like the beverage industry, it is also required to demineralize the water. The systems used for demineralizing utilize multibed or mixed-bed systems. Mixed-bed systems are more compact and require less space while producing high-quality water. Multibed and mixed-bed ion exchangers are sometimes used in sequence to produce high-quality demineralized water.

The mixed-bed system works well on water with limited alkalinity or acidity caused by mineral acids. It will produce water with low conductivity. When the raw water has high alkalinity or mineral acidity, high-purity water can be obtained with a two-bed followed by a mixed-bed system. The water passes through a cation exchanger where the carbonates are converted to carbon dioxide that is removed in a decarbonator. In the anion exchanger, anionic impurities are removed. For final purification, the water passes through a mixed-bed ion exchanger. Mixed-bed exchangers have both cation and anion exchange resins in the same vessel. Here, most of the dissolved, ionizable solids will be removed.

The removal of silica (SiO_2) in a demineralizing system depends upon the use of a strong base anion exchanger.

11.4.4 Filtering

Filtration is normally the first step in water purification. It usually precedes softening and demineralization. In the case of microfiltration, it is frequently the final step in the system (Table 11.6).

Particles with diameters larger than 0.2 μm can be removed with membrane filters. Such particles will include solids, colloids, precipitates, cell fragments, algae, yeasts, and some bacteria. In many cases, the components that are to be removed can be attached to other particles that will be large enough to be removed.

When suspended solids are to be removed by membrane filtration, any load higher than 200 mg/L should be pretreated. The filters will be able to handle the load but will have to be cleaned so frequently that it will not be economical.

Table 11.6   General Operating Conditions for Membrane Filtration Systems

Contaminants	Feed	Filtrate	Removal Efficiency
Suspended solids	200 mg/L	<1 mg/L	99.5%
Colloidal particles	200 mg/L	<1 mg/L	99.5%
Turbidity	500 ntu	<0.2 ntu	99.96%
Silt density index	Over 5	1	N/A
Total coliforms	10^6 cfo[a]/100 mL	10 cfo/100 mL	10^5
Fecal bacteria	10^6 cfo/100 mL	1 cfo/100 mL	10^6

[a] cfo = colony-forming organism.

11.4.5 Care of Filters and Ion Exchangers

Membrane filters and ion exchange resins require a fair investment and should be maintained properly to ensure optimal usage before replacement. Strict adherence to the manufacturer's operating conditions will ensure trouble-free operation. Resin manufacturers will suggest procedures that will avoid fouling of the resins. The resins and filters are designed to remove dissolved and suspended solids and they stand a good chance of becoming fouled by the materials that they are supposed to handle. Once they are no longer effective they need to be cleaned or regenerated according to instruction to be effective.

If a cation resin that has many calcium ions bound to it is subjected to regeneration with sulfuric acid, the acid will react to form insoluble calcium sulfate. This will foul up the resin. It is very important to make sure that the acids are applied in the right sequence when the cleaning is done. Normally, low flow rates, temperature above 30°C, and sulfuric acid concentrations over 5 percent will encourage the formation and precipitation of calcium sulfate.

Once calcium sulfate (plaster of Paris or gypsum) precipitation occurs, the problem can be remedied by backwashing with hydrochloric acid. Remember that sulfuric acid is a stronger acid than hydrochloric acid.* Backwashing the bed with large amounts of soft water is the first step. The "loose" calcium sulfate can be removed in this way. The resin is next treated with 1.2 mol/L hydrochloric acid. The amount of solution required is about 1000 L per cubic meter of resin at a flow rate of 70 L/min/m³. Formaldehyde can be added to stop corrosion of metal surfaces. It is best to avoid the formation of calcium sulfate. Whatever the remedy that is

---

* Sulfuric acid will react with a chloride to form a sulfate and hydrochloric acid. The reverse is not possible.

used, it will normally only be partially successful. After an enormous expenditure in chemicals and effort, it may still be advisable to replace the resin.

Silica can hydrolyze to silicic acid that can form polymers of colloidal silicic acid within the beads of resin. In this case, the bed can be restored with an application of a warm solution of sodium hydroxide. The caustic solution should be prepared with soft water and in severe cases it can be left in the exchanger for 18 hours. Consult the supplier of the resin before attempting this procedure.

After any treatment of the resin bed, the bed should be rinsed thoroughly and then regenerated before returning it to service. If fouling occurred, investigate the cause of fouling and ensure that it is avoided in the future.

11.5 Treatment of Boiler Feedwater

Once the normal water treatment is done to remove calcium and magnesium hardness, iron, copper, colloidal silica, and other contaminants, the water is ready for feedwater treatment. The first treatment will eliminate corrosive gases. Dissolved oxygen and carbon dioxide are among the principal causes of corrosion in the boiler. The corrosion caused by the gases is frequently the lesser problem; deposition of metal oxides in the boiler is a bigger problem.

Products of corrosion will become concentrated within the boiler, especially in areas where the heat transfer is high. The metal oxides will cause resistance to heat transfer in the most vulnerable part of the system. This can lead to local overheating and failure of the component. The deposits will also become heavier over time, causing pipe restriction and reduced circulation. The easiest way to deal with the problem is to avoid it; get rid of the oxygen and the carbon dioxide in the feedwater.

11.5.1 Deaeration

The removal of oxygen, carbon dioxide, and other gases from feedwater can be accomplished in more than one way. The feedwater can be heated to reduce the solubility of the gases. This will also increase the efficiency of the boiler.

Small amounts of oxygen in feedwater can cause localized pitting that can cause boiler failure even when only a small amount of corrosion took place. Boilers are constructed from carbon steel, and water is the heat transfer medium. The potential for corrosion is therefore high. Iron in feedwater is normally in the form of an oxide. The two types of oxides are the red iron oxide (Fe_2O_3) and the black magnetic oxide (Fe_3O_4). The red oxide is formed in oxidizing conditions while the black oxide is formed under reducing conditions. Red iron oxide is converted to an insoluble hydroxide as soon as it gets into the high-temperature, high-alkalinity area of the boiler.

Water will rapidly corrode mild steel. The Q10 rule states that chemical reactions will double with every 10°C change in temperature. At boiler temperatures, the reaction rates are fast.

$$3\,Fe + 4\,H_2O \rightarrow Fe_3O_4 + 4\,H_2$$

The black iron oxide or magnetite is a normal product of corrosion. In new boilers, this reaction will take, and a film of magnetite will cover the surface of the boiler. This film of magnetite will inhibit further reaction and thus will protect the boilerplate. The layer will be about 0.025 mm in thickness when further oxidation is inhibited. Obviously, oxidation will still take place and a boiler will show about 1 mm of corrosion per year. This is one of the reasons why we use boilerplate that is much thicker than what is warranted when it is new.

Water is frequently deaerated in a system where a fine spray of water comes in contact with steam. The steam will heat the droplets sufficiently to remove the dissolved gases. The gas is then vented and the deaerated water is ready for use.

11.6 Wastewater Treatment

In big cities, wastewater will be discharged into a big body of surface water. In areas where land is cheap, wastewater can be disposed in surface irrigation systems. Whatever the means of disposal, the wastewater should not affect the ecological balance of the large body of water. Even very large bodies of water, like the Mediterranean Sea, are so polluted that the ecology has been destroyed in many areas.

To preserve the ecology of the natural water body, the wastewater must be treated. Any substance or organism in the wastewater that will pose a health risk to the public must be removed, destroyed, or reacted to make it harmless. In the same way, any substance or organism that could pose a threat to the ecosystem must be rendered harmless before the wastewater is introduced into the surface water. Suspended solids and particulates must be removed. Soluble solids must be diluted to a level that will not harm the ecosystem. Nitrates and phosphates must be removed to stop excessive proliferation of fungi in the water. All natural surface waters are public property and no one has the right to foul something that belongs to all of us.

The degree of wastewater treatment is dependent upon the local environmental conditions, the local and state regulations, and the federal regulations. There are standards for natural streams and there are standards for effluent streams. Stream standards are promulgated to prevent the deterioration of natural stream water quality. The standards set limits on the amounts of specific substances that are deemed pollutants for streams, rivers, and lakes. The standards include dissolved oxygen, coliforms, turbidity, acidity, and toxic substances.

The effluent that comes from a factory will be covered by effluent standards. In this case, the factors that need special attention will include biological oxygen demand (BOD), suspended solids, acidity, and coliforms.

Primary treatment of wastewater will be to remove most of the particulate material and about 60 percent of the suspended solids. This will cause a drop of about 30 percent in the BOD. Dissolved substances are not removed in primary treatment facilities. Primary treatment involves simple processes such as screening and perhaps centrifugation.

Secondary treatment will address the total BOD issue. In this case, the suspended solids and BOD will be reduced by about 85 percent. A minimum level of secondary treatment is required before wastewater can be released into the natural water system.

If the wastewater contains high levels of dissolved solids, nitrates, and phosphates, tertiary treatment must be done. Tertiary treatment can remove more than 99 percent of the impurities from water. The effluent after tertiary treatment is clear water. Tertiary treatment is expensive and will double the cost of secondary treatment. Industry will use it in very special circumstances.

The final step in any wastewater treatment is disinfection. Chlorination is normally used as a way to disinfect water. Since high levels of chlorination may have adverse effects on the environment, the residual chlorine is frequently removed before the water is released into a natural stream.

11.6.1 Water Conservation

The cost of fresh, potable water is frequently lower than the cost of getting rid of the used water. Water conservation saves costs at both ends of the equation: paying for less water as a commodity and having a smaller volume of water to dispose of. The sanitation of the operation and product safety should never be compromised in a water reuse operation.

Every plant should map water usage and then decide where water can be moved from one process and used in another process. In some cases, water can be reused without any treatment. In other operations, water might require treatment before it can be used somewhere else.

Water that is used as a final rinse for clean equipment can be reused as the initial cleaning water on some other process. The final wash water from product lines can safely be used as initial wash or fluming water for the same raw material. The countercurrent principle is very important. Water used for can cooling can often be saved and used for fluming raw products if it does not contain toxic rust inhibitors. Every possible precaution must be used to avoid contamination of any product with used water.

Reuse of washer and flume water (using the counterflow system) must be carefully controlled to prevent excessive accumulation of soil and organic debris.

Reclaimed water for washers and flumes should be chlorinated before reuse if the temperature is higher than 27°C.

All water for reuse should be screened to reduce solid buildup. Chlorination is recommended for all reused water and the residual chlorine level should be increased to 4 mg/L twice a month to make sure that organisms that could be harbored in the organic-rich environments are killed.

11.6.2 Fluming Debris

Food processing plants use large volumes of water to process food products and to clean plant equipment. This yields large amounts of wastewater that must be treated. Using more water than what is essential will produce excessive volumes of wastewater that add financial and ecological burdens to the processing plant and to the environment. There are many ways to reduce water use and wastewater production. This will also eliminate many problems and costs associated with wastewater.

When water is used to wash processing debris into the floor drains, a large amount of blood, loose meat, loose pieces of food, soluble protein and sugar, and inorganic particles enters the wastewater stream. Most of the waste will increase the BOD of the effluent considerably. Wastewater treatment plants use BOD levels to gauge the amount of waste that is present in water. Higher BOD levels require longer treatment and sewer plants charge on the basis of BOD level in the water.

Depending on the size of operation and the type of effluent, water treatment can cost as much as $1,000,000 per year. The public is very environmentally conscious and this should be an additional motivation for industry to be more responsive to environmental issues.

11.7 Treatment Facilities

The Environmental Protection Agency is particularly vigilant about our water supply and the ecosystem connected to effluent systems. No industrial effluent is allowed to have a negative impact on the environment. This includes the alteration of the pH or temperature of the natural stream into which the effluent flows. Another important issue is the biological oxygen demand (BOD) of the effluent. If the effluent contains large amounts of nutrients that will benefit a particular natural organism, that organism will grow at a very rapid rate and deplete the oxygen in the water, causing the demise of other organisms, including frogs and fish.

Industrial water treatment can be done on the premises or the water can be channeled into municipal treatment plants, where it will be treated at a price. The price depends upon the average load, the type of organic content, the maximum load, and the volume of the effluent. Many plants are spending large amounts of money to pretreat effluents to reduce treatment costs in municipal facilities.

11.7.1 Aerated Lagoon

An aerated lagoon consists of a large pond or tank equipped with mechanical aerators to maintain an aerobic environment. Aeration also keeps the organic materials suspended. When it is allowed to settle, facultative bacteria can rapidly deplete the oxygen supply in the sediment and allow anaerobic organisms to start growing. This will give rise to all the horrible smells that are frequently associated with treatment lagoons.

Aerobic digestion results in complete biological oxidation leading to breakdown product with low or no odors. The breakdown in an aerated lagoon is slower than in an activated sludge system, but it allows the appropriate amount of time for the breakdown of complex materials. The microbial population is also less sensitive to changes in effluent strength and types. The effluent from the lagoon is normally channeled to settling tanks to remove suspended solids.

11.7.2 Trickling Filters

In trickling filters, use is made of bacteria that grow on some medium over which the water is percolated until the desired breakdown of material occurs. The wastewater is first placed in settling tanks to remove suspended solids. The water is then pumped to the trickling filters, where it is distributed over the medium with spray nozzles. The bacteria are exposed to the medium and air, so a strongly aerobic system exists. Trickling filters are used to treat wastewaters with poor settling characteristics.

11.7.3 Activated Sludge

The activated sludge process is frequently used as the principal process in municipal water treatment facilities. The activated sludge system consists of a storage tank to equalize the wastewater, a settling tank, an aeration tank, and a clarifier. It is important that the wastewater be mixed in the equalization tank. Bacteria in activated sludge systems are sensitive to abrupt changes in the water composition.

After the solids have been removed, the water enters the aeration tank where an aerobic bacterial population is maintained. Oxygen is supplied to the aeration tank by pumping air through a distribution system. The effluent from the tank goes through a clarifier where the biomass and water are separated. Some of the biomass is recycled as activated sludge while the rest is dewatered and dried.

11.7.4 Anaerobic Treatment

Anaerobic biological treatment starts by settling the sludge. The supernatant is pumped to a digester where the organic material is fermented by anaerobic bacteria. The breakdown products are varied and the final electron acceptors are organic molecules. This can lead to the formation of foul smelling substances if the waste-water contains significant amounts of sulfur compounds.

11.8 Waste Disposal

When most of the waste was flumed down the drain, the problems were economical or at least were not part of the solid-waste problems. Conservation of water causes more solid waste and this increases the burden on solid-waste disposal.

Solid-waste disposal is expensive and has many regulations associated with its proper disposal. The kind of solid waste that is generated is the first item that should be considered. This information will be used to select the best disposal method in accordance with regulations. Other considerations are the amounts of the different types of waste, the seasonality of certain waste products, and the cost of different ways of disposal. Many plants have been moved to other locations because of waste disposal problems.

The quantity of waste can be predicted with a fair degree of accuracy. Since this can be determined in the plant planning stage, the proper handling of waste and its disposal will be part of the overall plan. In many instances, new lines are introduced into existing plants and the waste handling is not always given the consideration that it deserves.

11.8.1 Prevention of Waste

The best way to deal with waste problems is to minimize the amount of waste that is generated. Frequently, one or more of the waste products could be used as a raw material for some other product. In most meat packing plants, the waste has been eliminated and changed into raw materials for someone else. Even the hair inside cattle's ears is shaved and used for art brushes. A standard joke is that only the squeal is lost.

To reduce the amount of waste, the following could be considered:

1. Do not flush possible by-products into the sewer.
2. Use refrigeration to prevent spoilage.
3. Process by-products into a stable product as soon as possible
4. Collect usable product and reprocess it.
5. Set up a waste prevention program.
6. Start a recycling program for packaging and paper.
7. Determine if one or more of the waste streams could be used somewhere else.

Spent grain was an enormous problem in the beer industry until its value as animal feed was "discovered."

11.8.2 Solid Wastes

The solid waste that is generated in the food industry includes sand, grit, peels, trimmings, pulp/skins, hulls, and stalks. Collection and transfer of the materials is a materials handling problem. Collection and handling must be efficient and sanitary. Full bins with flies above them are not allowed in a food plant. Storage of the waste prior to disposal is another area that can be problematical. Many insects love wet pulp as a breeding ground. The wet pulp will also sustain an active microbiological population with its accompanying odors and decomposition products. Broken grain can attract all sorts of rodent and bird pests that will become a problem in warehouses in a short time. Some waste, like sugar beet pulp, can be used as animal feed or it can be tilled into the soil to increase the organic material in the soil.

Solid waste from meat and fish packing operations should be processed as soon as possible if it has economic value or changed into something that will be stable until it can be disposed of. In the fish industry, all the guts, scales, and skins are dissolved in alkali. The solution is then neutralized and emulsified as an organic fertilizer. Leaving fish waste unprocessed for only a short time leads to so many odor complaints from everyone in the area that it is just not worth it to draw so much negative attention to your operation. When temporary storage of putrefactive products is necessary, storage containers must be watertight, corrosion resistant, and easily cleaned. The containers in which putrefactive wastes are stored must be cleaned daily. High-pressure hoses and steam cleaning will normally keep the whole system clean and sanitary.

Solid waste can also be compacted to minimize the space that it occupies. Compaction or screening may be used, in order to reduce the volume, and a suitable drain and flushing system must be used to prevent flooding of the collection area.

11.8.3 Waste Incineration

Incineration is the controlled high-temperature destruction of solid-waste materials to dispose of the solid waste, dispose of hazardous waste, and produce energy from waste. Many companies use waste as part of the fuel in boilers or processes. Waste is incinerated in a double system. The first section is the burner that can be fueled with gas or oil to start the incineration process. The solid materials will be combusted in this part. All the gaseous compounds will move to an afterburner, where they will be further oxidized.

Since food plant waste is organic in nature, it can be assumed that the main gaseous products of incineration will be water and carbon dioxide:

$$C_aH_bO_cN_dS_e + nO_2 \rightarrow CO_2 + H_2O + SO_2 + NO_2$$

Since the amount of nitrogen and sulfur is small in comparison to carbon, oxygen, and hydrogen, the amount of sulfur dioxide and nitrogen oxide will also be limited. The ash will contain phosphorous compounds, metal oxides, and waste that did not burn.

11.8.3.1 Controlled Incineration

Incineration is not just burning rubbish. Good control will allow control of emissions and therefore compliance with regulations.

The temperature in the incinerator must be selected according to the waste that is handled. At temperatures below 800°C, combustion will be incomplete and soot will result. Products such as dioxins and dibenzofurans can also be formed. At temperatures from 900°C to 1100°C, all hydrocarbons will be destroyed.

The residence time of material in the incinerator is very important. Short residence times can lead to incomplete incineration and air pollution. The longer the material is held in the high temperature zone, the more complete the incineration is. Some solids require residence times of several minutes.

Excess oxygen will ensure complete oxidation of compounds, producing fully oxidized compounds with little environmental effect. With an oxygen concentration of 150 to 200 percent of the calculated amount for the specific waste, oxidation will dominate and by-products of incineration will be kept to a minimum. Turbulence in the incinerator enables good mixing of waste and air.

11.8.3.2 Incinerator Emissions

All incinerators emit gas and particulate material. Atmospheric emissions include carbon dioxide, water, oxides of nitrogen, sulfur and phosphorous, and halogen acids. Sulfur oxides will form acids when they dissolve in water.

The acidic gases, sulfur dioxide and hydrochloric acid, are the biggest problem from an environmental point of view. They are formed in the incinerator when compounds containing sulfur and chlorine are burned. Both gases are water soluble and they can be removed from the effluent gas streams with water sprays (scrubbers). When alkaline solutions are used as sprays, the acids will be neutralized when they are removed from the gas effluent.

Some of the small particles of ash and raw material can be carried out of the furnace with the gaseous emissions. The scrubbers will normally also remove these fine particulates. In most cases the cleaned gas is not warm enough to move up the smoke stack and some reheat system is required in the flue.

11.8.4 Landfill

Landfill is a popular method of waste processing. It must be carefully controlled to eliminate insect breeding, odor problems, and groundwater contamination. In a landfill, it is important that the layer of waste must be covered with soil within a predetermined time. It requires a large amount of waste before landfill is an economical method of disposal for one factory.

Sanitary landfill is an area of specialization that few people recognize as a science. It requires planning, careful construction, and control. In the landfill the solid waste is spread in a thin layer over the total area. It is then compacted to a minimum volume and soil is spread over it to minimize environmental problems.

The operation of the landfill does not stop when no more waste is added. From that time, it is still controlled to ensure that some ecological disaster will not happen while digestion carries on. Landfills are normally monitored for about 15 years after closing. Development of marsh gas is not uncommon on old landfills and can be rather dangerous if it is not vented properly.

In many cases cities are finding great resentment from communities living in areas close to new proposed landfill sites. There are concerns about impact on property values, public safety, and odor and emission control.

11.8.4.1 Landfill Methods

There are three common methods for sanitary landfill: the trench method, the area method, and the canyon method.

- The trench method utilizes a trench that is excavated for the purpose of waste disposal. The excavated soil is stockpiled for use in the soil that must be spread over the waste. Sometimes regulations will specify that a trench be lined. The waste is spread over the area and then tamped down before it is covered with a layer of soil. This will continue until the landfill is about 3 m in depth. Once this height is reached, a thick layer of soil is applied and an adjacent area is used for the landfill until it also reaches 3 m. At this point, more waste is piled on top of the original area and it is built up in stages over many years. Once a height of 50 m is reached, the landfill is closed with about 1 m of soil. The final cover is domed to allow quick drainage of water away from the landfill. This will reduce seepage from the landfill and possible contamination of natural water.
- The area method is used where trenches cannot be used due to the type of soil. The operation is similar to that of the trench method with the exception that it is all above the natural ground level and soil must be hauled in.
- The canyon method utilizes natural and man-made depressions for landfill purposes. Abandoned quarry sites are frequently used. The operation is the same as that used for the trench method.

11.8.4.2 Public Health and Environment

A landfill must operate in such a way in which it will not detract from the environment. The site should be screened from roads and residents by planting and landscaping. Waste should not be allowed to blow around in strong wind. Any water that leaches from the landfill could be heavily contaminated with any soluble substances that were generated during the breakdown process.

Birds are attracted to landfills and can become a serious problem. No landfill is allowed close to an airport because of the danger to aircraft. The best way to deal with odor at a landfill is daily covering and sufficient compacting of the landfill. Gas generated during decomposition should be controlled. In March 1986 a house in Loscoe, Derbishire, was completely destroyed by a methane gas explosion. The occupants of the house were injured but survived. The source of the methane was the local landfill. An atmospheric low-pressure system caused the air pressure in the house to be lower than the pressure of the gas in the ground. The gas flowed into the house and, when the central heating furnace started up at 6:30 a.m., the gas exploded.

Chapter 12

Materials Handling

12.1 Importance of Materials Handling

Materials handling is defined as the organized movement of a specific material from one place to another, at the right time and in the right quantity. It may involve lifting, moving horizontally or vertically, and storage of materials. It is an important factor in the smooth operation of any food processing plant and it can greatly affect the quality of the product, as well as the cost and profitability of the operation.

From the time that a material enters the factory gate, cost is added every time it is handled. Each movement also carries the risk of damaging the material. Material handling is so commonplace that it is frequently overlooked in cost calculations. Everybody knows that shipping should be included as a cost. Material handling and flow should be seen as an internal shipping operation that has an associated monetary value. The cost of material handling, storage, and movement must be added to the final product cost.

In many operations in which production has been streamlined to be as efficient as possible, little or nothing has been done to make materials handling more efficient. It is no wonder that materials handling costs can be as much as 50 percent of the total manufacturing cost. Improper materials handling can be a major factor in waste and product damage. Every time a material is moved, there is danger of bumping, dropping, and crushing. Unsafe operation of industrial trucks, conveyors, and carts has been responsible for many serious plant accidents.

Materials handling affects the cost of manufacturing, distribution, and selling price of all food products. Modern materials handling methods are directed at accomplishing movement and distribution with a minimum of labor, waste, and cost, within the shortest time and with maximum safety.

In the food industry, materials handling is concerned with the movement of raw material from a warehouse, supplier, or the receiving bay to the bulk storage area or to a processing line. During processing, materials need to be moved from one process to another and, after processing, the material needs to be moved from packaging to a warehouse or to dispatch.

Setting up a materials handling plan requires a fair amount of effort and the expectation is that it will yield concomitant benefits. The objectives of materials handling include the following:

1. Decrease in the handling cost by better utilization of labor, machines, and space
2. Decrease in the operational cost
3. Reduction of production or processing time
4. Increase in efficient use of storage space
5. Keeping material moving, thereby reducing the space occupied by in-production material
6. Preventing handling-related injuries and accidents
7. Improvement in product quality
8. Reduction in material wastage

12.2 Increasing Materials Handling Efficiency

It is important to study the basic materials handling system. A system is an organized group of individual objects and actions that must operate in harmony to accomplish a purpose. It is important to match the sizes of the lugs, bins, or boxes that have to be moved to the size of the pallets, which are dictated by the width of the aisles. The transport cart must be able to handle the pallet size and still maneuver around tight corners.

In most cases, all the objects and actions will be in place; all that is required is to manage the system so that all the subsystems will work together to deliver the right material at the right time to the right place in the right quantity and condition. To make the system more efficient, the following can be used as guidelines:

1. Minimize movement. Shorten transfer lines as much as possible and eliminate all nonessential material movement.
2. Handle products in as large units as practical or as unit loads.
3. Remove as much water from product that needs shipping and storage. It is not economically sound to ship or store water if it can be avoided.
4. Change the form of the product for easy handling (bag in box for liquids).
5. Change the operation to continuous flow if possible.
6. Automate materials handling as much as possible.
7. Use an appropriate production management style (just in time [JIT], kanban, etc.).

Once the materials handling system has been scrutinized, it is important to look at each individual operation in the subsystems. The following questions should be asked:

1. Is the operation essential? Can it be streamlined for greater efficiency?
2. Is the cost satisfactory? Can it be reduced?
3. Are units sized for maximum efficiency?
4. In what way does the handling operation or system impact the final product quality?
5. Can the operation be improved?
6. What suggestions did the operators make?
7. Can the operation be done in a shorter time?
8. How will mechanical systems impact a manual operation?
9. Is it possible to automate the system?

Mechanization is much cheaper than manual labor. A person that works hard can generate about 70 W per hour. At minimum wage of $6 per hour, this labor will cost

$$\$6 \times \frac{1000}{70 \text{ W}} = \$86/\text{kWh}$$

If the work can be done with an electric motor, the same kilowatt-hour in electrical energy will cost less than $0.10. Use a mind where it is required. For mindless tasks, use something without a brain or use a brain to control many mindless machines at the same time.

12.2.1 Designing a Materials Handling System

Each subsystem of the materials handling system should be analyzed in isolation. The boundary should then be enlarged to include associated subsystems to analyze the balance between the subsystems. By stepwise enlargement of the boundaries, the whole materials handling system will eventually be within the boundary.

At this stage, the boundary should enlarge to include parts of the process affected by the materials handling system. At this point, the analysis should show the movement of all incoming raw product and ingredients and their movement into storage. It should also show the major product flow and all specialized or branch movements, the various processing operations, and the final product storage. Finally, the movement of finished goods should be considered and analyzed. It is essential to note all high-traffic areas and the sequence of movements so that all can be itemized. Analyses of these data will be used to streamline materials handling for the entire operation.

The simplest design that will do the job is usually the easiest for workers to understand, the cheapest option, and the most dependable system. Other considerations that have to be taken into account when components of a system are chosen will include ease of lubrication, maintenance cost, and corrosion resistance. The speed control regulators, safety devices, and switches should also be chosen to ensure versatility and durability.

12.2.2 Kinds of Systems

In a production operation, materials need to be moved to different processes. Food products are frequently pumped through the factory. The system is efficient and uses much less water than the fluming that used to be the mode of transport, and it can handle variable speeds.

Product should be robust enough to tolerate the pump action. Peach halves are pumped with centrifugal pumps without damage to the product. Pumping whole ripe tomatoes might leave you with a thin tomato juice.

When slurries and viscous products are pumped, the pump must be chosen so that it will suit the characteristics of the products. Shear thickening materials, such as heavy starch slurries, should not be pumped with a centrifugal pump. At the right amount of agitation, the starch slurry will turn into a solid. As soon as the agitation stops, it will flow again. Unrefined fish oil will do the same.

In some cases, the pumping action is used as the force to operate a unit operation. In the french fry lines, the potatoes are pumped from one area to the next. Close to the exit of the pipe, the knives are installed that will cut the potatoes into fries. The pump water will wash away most of the loose starch, which will be recovered in a centrifugal clarifier.

Pumping dry solids is used extensively in the milling industry. Pneumatic systems are also used in processing and packing lines for various dry mixes. It is frequently the most efficient transfer system, but only suitable in certain operations.

Conveyors can handle anything from lumps of coal to the most delicate berry fruits. They are available in many sizes and types and have many fitted devices to assist in material flow and orientation.

12.2.3 Analysis of a Materials Handling System

When any decision has to be made regarding a change in materials handling, one should start by looking at the present system. Determine the present handling cost per unit production. List some requirements that are not met by the system as it exists and define what will improve the system. Check how much the new system or part of the system will cost. Determine the new handling cost per unit produced.

Table 12.1   Checklist Regarding Materials Handling Inspection

Incident	Description (who, what, where, how much?)	What action should be taken?
Manual lifting of materials		
Materials piling up		
Cluttered floors		
Unlabeled materials and products		
Unsafe handling practices		
Machine operators waiting for materials		
Trucks and trolleys waiting for materials		

Included in the calculations should be items such as the cost of demurrage.* The maintenance and repair costs that will be incurred on the present and future systems must also be considered. A calculation of the product cost that is lost because of bruising, falling from the transfer system, and contamination should also be made. When inspecting a production area, it is good to have some kind of checklist (see Table 12.1).

The receiving department is frequently a good example of every possible bad materials handling habit. The following aspects should be inspected:

1. The materials and loads that are unloaded by hand
2. The amount of materials that are standing around in the reception area before they are moved into a storeroom
3. Temperature-sensitive materials standing around in reception bays—one of the big problems with frozen products at retail stores
4. The number of people standing around
5. Movement of materials to make sure that they are moved as little as possible and along the shortest route to the storeroom

Materials handling problems affect all departments. Within the production area of the plant, one must make sure that skilled workers do not handle heavy

---

* Demurrage is the compensation paid for delaying a ship, freight car, or truck at the loading point because of inability to load in a set time. When 24 trucks are delayed 1 hour, the value of one truck's transport for a day is lost.

materials. It is a waste of money to allow someone that is paid at a higher rate do the work of a laborer.

Determine the frequency with which delicate materials arrive damaged at the production point. Once a product is produced, how fast is it removed from the production area? Clutter makes good performance difficult. Check that every pallet load or tote box is clearly labeled.

Check what happens to scrap and damaged products. They should be removed from production by a scrap disposal system. Aisles should be well lighted and kept clear.

Within any industry, materials handling starts and stops at the reception and loading bays. There are storage areas that keep the loading bay stoked and the reception bay empty. Inventory rotation and control are the heart of the entire materials handling system. The checks on the storage areas should include the following:

1. The level of light in the area should be sufficient to identify materials easily.
2. Storage areas should be demarcated into specific sections where materials are kept—liquids in one area, dry powders somewhere else, cleaning materials in their own area, etc.
3. Record keeping of the inventory should be automatic and continuous. At any point in time the amount of a specific material should be available.
4. The retrieval of material should be easy and accessible with the right equipment.
5. Loading limits of stacks and floors should be clearly displayed and storage areas should not be overloaded.
6. The materials should be stored in a form that is easily handled.
7. Is the material moved through the storage area or does it move in and out through the same door?
8. The aisles must be wide enough to allow smooth operation of trucks, carts, and lifts.
9. The storage area for a specific raw material should be as close as possible to the production area where the ingredient is required.

In the shipping department, the following points should be considered:

1. Simplifying and cost reduction when power conveyors or trucks are used
2. Ensuring that outgoing lots are ready when truck or railway cars arrive
3. Ensuring that the temperature of the product is maintained at prescribed levels while it is in the loading bay and during transfer to the trucks
4. Ensuring that shipments are not allowed to clutter the dispatch area if something happened to a scheduled truck
5. Clearly labeling all products waiting to be moved

Remember that the objective of a good materials handling system is to move the specified materials safely, at the lowest cost, with the highest efficiency, and with the least amount of damage to materials.

12.3 Materials Handling Equipment

There are many different pieces of materials handling equipment. Some are very sophisticated while others can be simple. One of the most sophisticated materials handling systems can be found in the music industry, where the total production of music CDs is automated. A robot will collect the master and place it into the copy machine to produce the required number of CDs; the robot will then take the master CD back to its place. With about 10,000,000 masters, this is quite a feat.

Materials handling equipment can be grouped in fixed-path, fixed-area, and variable-path–variable-area types. Fixed-path equipment can move material from one point to another. In the food industry, the conveyor belt systems that supply the empty containers, the sorting belts, and the chain-mesh belts going through continuous baking ovens are all examples of fixed-path conveyor systems.

A fixed-area materials handling system can serve any point within a specific volume. The overhead traveling crane is an example of this type of materials handling. The crane is used to handle heavy loads within large manufacturing areas or in warehouses. The gantry cranes at harbors run on tracks, making them operate in a fixed path. The area that they can handle is dictated by the tracks, so they also fall in the category of fixed-area cranes just like overhead cranes.

The variable-path–variable-area handling equipment will include all manual and motorized carts, forklifts, trolleys, and dollies. These units can be moved anywhere in a factory and perform whatever material movement they can perform.

12.3.1 Conveyors

Conveyors are units that will cause horizontal or inclined continuous movement of material. The system is normally fixed in place and conveying will take place in overhead, working-height, floor-level, or under-floor areas. Material motion is induced by gravity or mechanical means.

12.3.1.1 Chutes

Chutes are similar to a child's playground slide. They are smooth surfaced, inclined troughs that allow materials to move down the chute under the force of gravity. The chute can be made of steel, plastic, or wood. Since chutes are used to move materials down an incline, it is obvious that two forces are acting on it: gravity that will try to pull it straight down and a horizontal component. At a zero incline, gravity is not large enough to pull the material through the chute. As the incline becomes steeper, gravity can overcome inertia and allow horizontal movement with vertical movement. At the point when the chute incline is so steep that it is vertical, the material will drop under the force of gravity and friction will not slow it down at all.

The amount of friction is not very dependent upon the area of contact. When sliding a refrigerator across a floor, the frictional force will be the same if you slide it on its top or on its side. In fact, if four feet had a coefficient of friction that is the same as that of the sides, the frictional force will still be the same. The second frictional fact is that friction is proportional to load. If a second refrigerator were placed on top of the first one, the frictional force would double. The ratio of friction to load is constant.

$$\mu = \frac{f}{N}$$

When products are placed in the chute, friction must be overcome; otherwise, the material will just stay where it is. The force of static friction (f_s) between two parallel surfaces in contact will act in a direction that opposes the direction of the force that is trying to initiate movement:

$$f_s \leq \mu_s N$$

N is the normal force and μ_s is the coefficient of static friction. Static friction is zero for a body at rest and will achieve a maximum value at the point where motion starts. Once the force of static force has been overcome, a force of kinetic friction (f_k) exists that will act in a direction opposite to that of the movement:

$$f_k = \mu_k N$$

The coefficient of kinetic or sliding friction μ_k is used to calculate the size of the force. The coefficient of static friction is normally larger than the coefficient of sliding friction (Table 12.2).

Example 12.1

1. What direct force must be applied to a wooden box with a mass of 50 kg to start moving it over a wooden floor? The coefficient of static friction between the box and the floor is 0.6.

$$f_s = \mu_s N = \mu_s \times mg = 0.6 \times 50 \text{ kg} \times 9.8 \text{ m/s}$$

$$f_s = 294 N$$

Table 12.2 Values for Coefficients of Static and Sliding Friction for Different Surfaces

Friction between Surfaces Composed of:	μ_s	μ_k
Rubber on dry concrete	1.9	0.9
Rubber on wet concrete	0.8	0.6
Steel on steel (dry)	0.8	0.5
Steel on steel (lubricated)	0.12	0.07
Lubricated ball bearings	0.01	0.01
Steel on Teflon	0.05	0.05
Steel on aluminum	0.6	0.5
Aluminum on aluminum	1.9	1.5

The crate must start to move so that the direct force (F) must be greater than 294 N.
2. If the force stays the same and the kinetic friction between the crate and the floor is 0.4, what will the acceleration be?
   The crate moves and there is a reduction in friction; the excess force can be used for acceleration.

$$F - f_k = F - \mu_k N = ma$$

$$a = \frac{F - \mu_k N}{m} = \frac{F - \mu_k \times mg}{m} = \frac{294 \text{ N} = 0.4 \times 50 \text{ kg} \times 9.8 \text{ m/s}}{50 \text{ kg}}$$

$$a = 1.96 \text{ m/s}$$

Remember that it requires energy to keep the object moving.

When bulk material is handled, the friction between the material and the chute as well as interparticle friction should be considered. Moving materials down a chute without damage is used to convey objects from high to lower levels under the force of gravity. The following factors should be considered during the design of chutes:

1. Friction should be overcome before any movement of material will take place. With bulk material, one should consider both the friction between the material and the chute and interparticulate friction within the material.
2. Moisture content of the material that is conveyed and the humidity affect kinetic and interparticulate friction.
3. The inclination of the chute affects the acceleration rates.
4. The length of the chute will determine the terminal velocity.

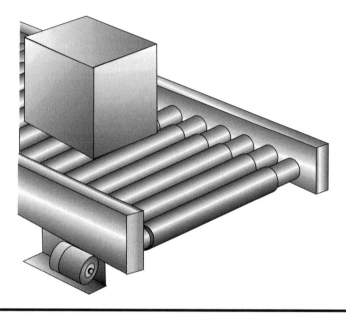

Figure 12.1   Roller conveyor.

12.3.1.2 Roller Conveyors

Roller conveyors can be powered or they may be free running (Figure 12.1). In the free-running conveyors, the rollers are mounted horizontally in a frame. This forms an almost continuous surface over which objects can be pushed. When the conveyor is inclined slightly (2° – 3°), the object will be moved by gravity.

Roller conveyors are used for handling items with firm bases like crates, cartons, and drums. At least three rollers should be in contact with the base of the item being moved over the rollers. Care must be taken when items like bags are handled on roller conveyors. If a bag is not very full, part of it might fall through the conveyor and cause a stoppage.

Roller conveyors can also be belt or chain driven. These powered rollers are used for moving solid objects horizontally or up and down through slight inclines. The conveying speed is in the order of 0.2 – 0.3 m/s.

12.3.1.3 Belt Conveyors

Belt conveyors comprise an endless belt that can be any width and any length. The belt is driven on one side, with an idler roller on the other side. Sheet metal or metal rollers are used to support the belts. The belt material can be plain or rubber-coated canvas, wire mesh, or stainless steel ribbon. The belts can have the sides supported at an angle to form a gutter. Smooth belts can be used for inclines up to about 20°.

If the belt is fitted with special slats or other devices to prevent rollback or slipping, the belts can be used on inclines up to 45°.

Belt conveyors made of special white polymers and running at working height are frequently used as inspection tables in the fruit canning industry. In the baking industry, the continuous ovens normally incorporate a chain-mesh or steel belt to move the product through the oven.

12.3.1.4 Slat Conveyors

Slat conveyors comprise metal or polymer slats that are attached to endless chain loops similar to belt conveyors. The conveyors can be made in different widths. In the lumber industry, the conveyors can be 6 m wide, while the smaller slat conveyors used in a canning industry will be as wide as one can. Many large manufacturers use floor-level slat conveyors for assembling machinery.

In the meat industry, the slats are in the form of large trays. The entrails, in the trays, move synchronously with the carcass hanging on a chain-driven rail until inspection is completed. In other industries where draining is required, the slats can be perforated.

12.3.1.5 Chain Conveyors

In the meat industry, the carcasses are suspended from hooks that are chain driven. The system uses a single drive chain that runs in a double I-beam system. The chain will latch to a unit that has a hook or basket suspended from it. The system is similar to a railway track with switches that can switch the buckets and hooks from one chain system to another.

12.3.1.6 Vibratory Conveyors

These systems will convey material forward by a slow movement in the forward direction and a rapid pull in the backward direction. It operates on the principle of inertia. This is similar to the magic act of pulling the tablecloth from under the dinnerware. This "push and pull" principle is used to move material over screens.

12.3.1.7 Screw Conveyors

Screw or auger conveyors operate on the principle of helical screws rotating in a semicircular channel or a tube. The helical screws push the material and the direction is reversible. Screw conveyors in tubes can be used in any position, from vertical to horizontal. The conveyor will move material at a speed that is regulated by the rotation of the auger. Discharge rates can thus be very easily controlled and it is frequently used in auger feeders.

The bulk handling of sugar, grain, beans, and other powdery and particular substances is well known. In the Netherlands, the windmills use an auger system to lift water at very high rates but at a limited height of about 1.5 m.

12.3.1.8 Bucket Elevators

Bucket elevators are large-scale bulk handling units. With their high capacity, they are used to handle particulate, free-flowing substances such as sugar, beans, and cereals. The steel buckets are carried on an endless chain. The buckets are loaded at the bottom of the system and tipped out at the top. It is an efficient and high-capacity system.

12.3.2 Trucks

Fork trucks are the most popular piece of materials handling equipment. They can go almost anywhere and can be fitted with special attachments so that they can safely handle drums and oddly shaped material. The basic truck is very versatile and this causes misuse; a lot of capacity is never utilized on small operations when other equipment would have been more suitable.

The trucks can be powered by electric motors running off batteries. The batteries are very heavy and the trucks have to carry this dead weight. They are used for short-range movements on smooth and level surfaces. Electric units are popular in processing areas where exhaust fumes could be a contamination hazard. They are also fairly easy to maintain and battery exchange can be done very rapidly.

The battery service center should be managed in a way that will ensure efficient operation and long service of industrial batteries. The battery center should charge and maintain batteries in top operating condition. The center should be designed so that the batteries are handled as little as possible and as safely as possible. Standardizing of battery size will obviously make handling much easier. The OSHA safety regulations should be used when the battery service center is designed.

The basic considerations when selecting a location for the center are the availability of power, ventilation, and the distance trucks must travel to have batteries charged. It is best to have the center as close as possible to the place where the units are used.

Trucks can also be powered by internal combustion engines using gasoline, diesel, or LP gas. These trucks should not be used in areas where the ventilation is poor or where the exhaust fumes could pose a contamination risk. Diesel- and gasoline-powered units are normally used outside, while the units inside normally use LP gas. LP gas burns cleaner and the resulting exhaust fumes contain fewer contaminants. It will still use a large amount of oxygen and replace it with carbon dioxide, so the ventilation of the area where it is used is still important.

The trucks are fitted with wheels that will operate best on the surfaces that are used. Trucks can be fitted with steel, aluminum, polyurethane, solid rubber, and pneumatic rubber wheels. Pneumatic rubber wheels will protect floors but have poor oil resistance and lack maneuverability. Steel wheels has good load-bearing capacity, wear resistance, oil resistance, and maneuverability but will damage floors and cause noisy, bumpy rides.

12.3.3 Pallets

The pallet system is widely used in the food industry since it allows the operator to handle large quantities of products at a time. It is also used when placing and retrieving products in storage. In many large food handling systems, the product is brought to the storage warehouse on pallets, stored on pallets, retrieved on pallets, and finally delivered to the retail store on a pallet.

The standard double-deck wood pallet is the basic unit. There is also the expendable pallet that is built very inexpensively and is often discarded at the retail store level or the local distribution center because it costs more to return it than to replace it.

12.3.4 Bulk Handling

Bulk handling is the most economical and efficient way to handle large quantities of food products. Most raw materials are handled in bulk. The farmer brings his crop to the processing plant or elevator where the whole truck is weighed before and after offloading. The change in weight is considered as the weight of the crop. Even finished goods are shipped in bulk. Ingredients such as oil, sugar, syrups, and flour are shipped in road or railway tankers.

Loading and offloading in bulk handling systems are much faster and more efficient and therefore cheaper than handling individual units. The product is, however, handled in one batch and any contamination is to the whole batch. Other factors that should be considered are industrial sabotage, pilfering, and terrorism.

Handling of liquid products such as milk is normally done in road tankers with flow meters that will accurately measure the amount of milk that is pumped from the farmer's storage tank into the tanker. At the dairy, the milk is pumped through a meter again to determine the size of all the milk in the tanker. If milk from one farmer is contaminated, the whole tanker will be contaminated and all the milk is lost.

Liquids can also be handled in small tanks that are installed into a metal frame so that they can be handled with a forklift and transported on a flatbed. The tank frames are normally the size of a fraction of a shipping container so that they can be bolted to the standard container bolts.

Free-flowing solids are moved pneumatically or with augers into railway cars or semitrailers. These have sliders at the bottom that can be opened to drop the cargo into a reception system that is under the railway tracks or under the trailer. Very rapid emptying means a quick turnaround of the empty transport system.

12.4 Plant Layout for Materials Handling

The overall layout of the processing plant will dictate the design of the materials handling system and will affect the cost and efficiency of the handling system. In established operations, the problems regarding materials handling can be a nightmare. In new systems, it should be part of the planning from the start.

Product movements cost money and it must be kept to a minimum. In some older factories, it is impossible to operate the plant without moving materials over long distances or even back and forth. The most efficient materials handling is when raw materials are received at one end of the plant. The final product then emerges at the other end without backtracking or sidetracking. This ideal is not frequently realized and many compromises need to be made.

A careful study of the plant layout, storage spaces, and processing sequences will allow planning to reduce the travel of product, people, and handling equipment. This will increase material flow, reduce bottlenecks and stoppages, reduce unsafe situations and practices, and could even increase product quality.

12.4.1 Space Arrangements

To plan the best handling system, the process should be charted in detail. All the operations that are used for every product that is manufactured in the plant should be listed. The chart should start with all the raw materials for each product and carry through to the dispatch side. If materials are added from different storage areas, the distance from the storage areas should be noted. A list should also be made of each ingredient, where it originates, where it is going, its nature, and the amounts involved per hour, per day, per week, or per year.

Since many products are normally processed in any food plant, an ideal one side in–other side out system cannot be used. Some parts of the plant in vegetable processing may be used for as little as 3 weeks of the year. Many compromises need to be made and the planning can be more inclusive than just a materials handling plan. Other items that can be included will be lighting, location of services, evacuation routes, and emergency stations. In many cases, a simple rearrangement can reduce materials handling significantly.

The type of equipment that will be used in materials handling should also be considered. If forked trucks are used, the aisles must be wide enough for two trucks to pass one another. If this is impossible, then one-way routing should be considered, or only one truck should be allocated to the area. In any event, aisles should be wide enough for the truck to pass workers at a safe distance. Crossroads should be avoided and if there are any, the area should be designed for clear visibility in both directions.

Openings into the area should be high and wide enough to allow the movement of fully laden trucks. If it is the only entrance into the area, it is normally wise to make the openings large enough to take out redundant equipment and bring in replacement equipment. All the aisles should be well lit so that the truck operators can see the limiting clearances easily. The aisles must also be kept free of any obstacles or overhanging machinery.

12.5 Efficient Use of Materials Handling Equipment

Equipment must be well maintained and should only be used for the tasks that it is supposed to be used for. People are frequently tempted to try a do-it-yourself operation when something heavy and awkward needs to be moved. Using expensive machines to perform tasks for which they were not designed can lead to breakdowns that can be much more expensive than just renting the right equipment. Operators should be trained how to use the machines properly. All operations must be scheduled so that the materials handling equipment can be optimized.

It is important that equipment be fully utilized. Some equipment will, however, be so specialized that it will only be used in certain operations. Special handling accessories for trucks, conveyors, cranes, and special lifting or positioning devices are not in continuous use but must be available when they are required.

To control costs of equipment requires planning and scheduling. This will allow best utilization of equipment and scheduled maintenance when one or more units will have slack.

12.5.1 Maintenance

Preventive maintenance, that stops breakdowns before they occur, is essential for any mechanical system. This requires scheduled checks of materials handling equipment to replace worn parts, service motors, replace fuel and gas lines, lubricate bearings, and service or replace batteries.

Battery-powered units should be charged at the end of each shift or whenever the batteries needs charging. If the batteries need more regular charging, they might require replacement. Gasoline and LP gas tanks should be refilled at the end of each shift or at the end of the day. Fast wearing parts should be replaced on a basis determined by the log of the specific unit. Bearings must be lubricated but not to the extent that they are overlubricated. The maintenance shop should overhaul each piece of equipment at least once a year. If the working conditions are very hard, the equipment must be overhauled more regularly. The downtime for a unit must be planned so that it might be necessary to have one or more units extra to what is required on a daily basis.

Service charts and maintenance charts should be available for every unit. Daily or biweekly operations can be charted on a wall chart. This table should list dates across the top and individual machine numbers along the side. The person performing the maintenance should initial the chart to ensure that the maintenance was performed. When parts are replaced, each part should be noted on a card, which is kept in a record system. This will help to track the maintenance scheduling and preventative maintenance.

Chapter 13

Manufacturing Plant Design

13.1 Building Design

An industrial building provides an environmentally controlled space in which the processing function takes place. The function is the important aspect and the design must be subordinate to this function. The building is also a large capital investment and must be designed to require a minimum of maintenance, a convenient operational environment for workers, and the right environment for sanitation and hygiene. The building must obviously be in compliance with state building codes, and federal agency codes for safety and hygiene.

Building designers of processing facilities must be aware of the many constraints within which they should operate. The state and federal standards are minimum standards and a company may choose to adhere to a more stringent standard for safety and hygiene. The first basic guidelines that should be used are the legal aspects.

13.2 Legal Aspects

The law regarding structures will be different in different states. Each state has an office that will administer the different federal and state statutes.

13.2.1 Building Bylaws

The state will have specific building bylaws that will define allowable areas, the types of buildings that are allowed, structural limitations, and requirements for lighting, mechanical systems, electrical systems, and materials. These will be minimum specifications that could and frequently should be exceeded. Sometimes, counties and municipalities have their own building codes that must also be adhered to. Remember that conformance means conforming to everything, even the most limiting regulation.

13.2.2 Sanitation Codes

Sanitation codes will be associated with federal law and state statutes and bylaws. The regulations will establish minimum specifications for sanitary maintenance of processing facilities. For animal products, the USDA is the regulating body and for all else it will be the FDA. Statutes are specific to the products that are processed and the equipment that should be used. They also include specifications for floors, drains, walls, ceilings, water supply, vermin and pest prevention, refrigeration, and personnel facilities. The USDA has a plan approval service that will approve the proposed building after scrutiny of drawings and specifications.

13.2.3 OSHA

State government generally administers the Occupational, Safety and Health Act. The codes are concerned with the requirements of continued safety to plant personnel. The safety codes will be covered in Chapter 15.

13.2.4 EPA

The Environmental Protection Agency is a federal agency that is frequently administered by the state government. The EPA regulations cover details and advice about water sources and effluent. The codes are set to ensure that water quality is preserved and that no waste is dumped into natural water bodies or city sewers that will cause excessive turbidity, acidity, or alkalinity; high bacterial oxygen demand; or large amounts of settable solids. The act also covers regulations concerning atmospheric emissions from smokestacks and dust from processes such as winnowing and dry product handling.

13.3 Expansion

New products and technologies become available at a very high rate. This causes continuous improvements and changes in processing lines. The old style brick and concrete structures are being replaced with modular designs that can be altered on the same or an expanded footprint. Very few processing plants that are built today are intended to last for 100 years.

In many cases, installing a machine that has double the capacity of the original machine, but the same physical size, can double production. The space requirement for the operation therefore stays the same although the output doubles. Replacing machines has to be done in a way that will take as little time as possible without compromising plant sanitation and with limited affect on production output. To do this, aisles should be wide enough to accommodate the movement of plant. At least one large entrance into the processing area is required. In some facilities, an area above part of the plant is reserved for moving equipment. This will normally require the installation of a fixed area handling system.

The expansion of freezer rooms can be a big problem if part of the floor adjacent to the freezer is not insulated during initial construction. During the construction phase, the cost for additional floor insulation will be minimal and if it is never used, that will be no problem. If the freezer needs to be enlarged and the floor is not insulated, this will require a step inside the freezer or the removal of a heavy concrete slab and pouring a new slab over insulation. This will not be possible in an operational food plant. The plant will have to be shut down while this work, which can take a few days, is done.

The provision of floor drains in future processing areas must also be considered. Adding a drain or two is not good enough. The whole floor needs to be sloped so that water will run to the drain, which will be at the lowest spot for the area. This brings about a small additional cost during initial construction, but a large cost if it is done later.

13.4 Plant Location

In many cases new plants are built in rural communities, where local authorities provide special incentives. These can be in the form of free land, cheap subsidized mortgage, or a property tax break. It is always surprising to see just where very large companies with national and international distribution networks have their headquarters. Digi-Key and Arctic Cat are situated in a small town with 8,000 residents. Polaris is situated in a town with 2,000 inhabitants and Marvin Windows, with 3,000 employees, has its main factory in a town with 1,800 people. All these companies are within a 40-mile radius of each other and all have sales greater than $500 million. The closest city with more than 1,000,000 inhabitants is more than 300 miles away.

Whatever the reason for considering a specific town in which to build the plant, the site should be selected after careful consideration of many aspects:

1. A site with a gentle but even slope from front to rear will allow easy installation of loading platforms at the rear of the building. A steep slope should be

avoided since this will increase building costs. If the site is large enough and if codes allow it, a gently sloping site will allow for water treatment ponds behind the building. There will be no danger of seepage back into the building. It will also drain rainwater away from the building. Streets are normally constructed with storm drains and the only rainwater on the site will be from the rain that falls on the site.

2. The site should have a soil-bearing capacity of at least 14,500 kg/m^2. With special construction methods and sufficient foundation work, problems with low soil-bearing capacity can be overcome. Rock in the subsoil will increase the stability of the soil but at the same time it can increase the cost of installing any subsoil work tremendously. Any excavation will also be very expensive. Dynamite blasting in a built-up area is always the work of specialists.

3. Sometimes a site can be found outside the city limits that will allow connection to city-owned sewers and treatment plant. The will require extended pipelines and, if the topography is unfavorable, it might require pump stations. This is a large additional and ongoing cost. Plant wastewater containing significant amounts of organic matter can cause problems at the off-site sewage plant and treatment will be done at a premium cost. Discharge into rivers means increased costs for sewage disposal. Discharge into rivers is usually tightly controlled and space may be required for some treatment of wastewater on the site. Solid-waste handling can become a hygiene problem very rapidly and should be controlled.

4. Industry uses various forms of energy for heating, lighting, and power. The availability of electricity and gas and the possibility or frequency of disruptions should be considered since disruptions affect processing and storage facilities. Current and future needs should be assessed regarding electricity and gas distribution on the plant site.

5. Accessibility to state-maintained and snow-ploughed highways in winter is an obvious benefit. Distribution by tractor-trailer is the norm and the processing plant must be designed to allow rapid bulk loading of trailers. Other important considerations are traffic problems, weight restrictions, and low or narrow bridges. Accessibility to railway transport is another option that should be considered. Many rural elevators were forced to close down because the railways were no longer interested in servicing them with one-car loads. For larger industries, getting deliveries by trainload is the cheapest form of transport. Mills will frequently receive grain by trainload.

6. The size of an ideal lot depends upon the facility. One hectare (10,000 m^2) is normally adequate for a small to medium size plant. The lot should be large enough to permit expansion. A food processing plant has the uncanny ability to outgrow its site, causing many safety and hygiene problems.

7. A truck service building on the premises has many benefits and a lot that can accommodate a service building will be ideal. A three-bay service building

will be about 18 m long and should include lift hoists or a pit, wash rack, tire storage, ablution facilities, lockers, and fuel pumps. The facility should also include a weigh scale to ensure load compliance. Trucks should be fueled prior to weighing.

8. Before a site is chosen, make sure that there are no long-term plans for area development or road widening. A road widening can very easily remove most of a parking area. Certain rights-of-way will also not allow permanent structures to be built on part of the land. The state highway department and the local authority planning office should be able to give details of any forthcoming projects.

9. Neighboring plants need to be assessed to determine the impact of the new facility on them and their impact on the new facility. A real problem that occurred was when a yeast producer built a facility next to and upwind of a dairy that made cheese, yogurt, ice cream, and various milk products. The dairy had to install special filters on all the air intakes and had to maintain positive pressure in the whole facility. Obvious contaminants such as smoke, dust, ash, and solvent vapors from automobile paint shops are easily recognized. Bacterial contamination is difficult to recognize before it becomes a problem. Nearby landfills and neighboring plants with large areas of surface water during the rainy months can be a breeding ground for many problem insects. The storage on site of discarded machinery, boxes, crates, and rubbish will become a breeding place for rodents and insects, causing serious sanitation hazards.

10. The availability of personnel to work in the plant is a major concern. The cost and the qualifications of the labor source will determine the average wages paid. The availability of training centers to increase the qualifications or abilities of the labor pool is beneficial. The availability and cost of technical services to augment maintenance staff are important considerations. Many plants depend on contract services and the quality and response time to get assistance should also be considered.

11. Modern processing also requires high-speed computer linking ability. The capacity, speed, and uninterrupted service are of great importance to all manufacturers. This extends to marketing and consumer relations. Distribution networks and transportation rely on good coverage with satellite-based communication systems.

12. The location of the production site relative to the source of raw materials and the distance from the major distributions centers must be considered. Frequently, either the raw material or the finished product requires long-distance transport.

Many properties will lack one or more of these advantages and the cost incurred by these problems should be projected to calculate the additional yearly expenses caused by them. Only then will it be possible to compare sites on an equitable basis.

13.5 Planning the Building on the Site

1. Locate the building about 45 m back from right-of-way. This will allow some space surrounding the building even after a highway-widening project.
2. Traffic should be planned to ensure safe movement of all vehicles and pedestrians. Places where pedestrians need to cross roads should be avoided. Truck and automobile driveways should be segregated close to the entrance with clear distinction for visitors. Soiled vehicles that could pose a sanitation risk to the product should be routed in a different way to avoid contamination. Roads should be kept in good condition to avoid damage to products in transit.
3. Provide access by drive to all sides of the buildings if possible. This allows one-way circulation and will allow access of fire-fighting vehicles to where they are needed.
4. Rail sidings, parking lots for trucks and automobiles, trash collection areas, and "surplus" equipment* dumps can become a source of contamination and a breeding place for vermin. These areas should be kept groomed and drained to prevent contamination to food products by seepage.
5. Locate loading bays at the rear of the building. A gentle slope from the front to the rear will cause the floor of the building to be about 1 m above grade without cutting or filling.
6. Grade the site for natural drainage to ditches. In some areas, grading must be done so that all surface water will be channeled into a storm drain system. Surface water must never be allowed to enter the waste drain system since water treatment costs are also based on volume.
7. Provide access to truck building and weigh scale.
8. Locate employee and visitor parking at convenient entry points to the building.
9. Truck parking should be separate from the other parking areas and could be fenced.
10. Raising outside equipment about 20 cm above the pavement will prevent it from being used by rodents as a breeding place.
11. Landscaping and gardens contribute to the company's image. Good planning will ensure low initial costs and low maintenance costs. Small recreation areas outside the plant improve employee morale. Employees should eat in the space provided for them and should not take food outside. Dropped food attracts birds, rodents, and insects, and feeding stray cats or birds on site should be prohibited.
12. Around the base of buildings, a 1 m wide grass-free strip that is covered with gravel or stones is advisable. This controls weeds and is a good area for placing traps and bait.
13. The perimeter should be fenced for security reasons and to prevent children from entering the grounds. This is very important if accessible bait boxes are

---

* A term frequently used for junk that someone does not want to throw away.

placed outside the plant. Chain-link fences more than 2 m high are normally sufficient. The fences will catch pieces of paper and cleaning them must be part of the overall housekeeping on the site.

14. Orient the building so that its best features and sign face the busiest street.

13.6 The Structure

Many hours of many committees are used to decide what the "ideal" new food plant will look like. The final design is normally a series of compromises to do the best within a limited budget. Choices have to be made regarding the construction materials, the complexity of the building, and whether a single-story or multistory design will be best for the specific location and process.

Single-story designs have the following advantages over multistory designs:

1. Materials handling systems are simpler.
2. Supervision and deployment of staff is easier.
3. Simple, straight production lines are possible.
4. With a very heavy plant, the expense to ensure adequate load bearing is much less.
5. It is easy to expand the floor space or add additional buildings to integrate with the main process.
6. Changing processing equipment is much simpler because of direct access to the transport area.

Single-story designs also have some disadvantages compared to multistory designs:

1. The footprints of single-story buildings are very large.
2. Gravity flow is limited to the height of the ceiling. When equipment is placed on the roof, is that a second story?

Architects are normally employed to design a processing plant. They are very concerned with aesthetics and space relationships. The architect partners with different engineers to design the structural integrity of the building and all its facilities. Within this partnership, many beautiful buildings are created that are not always practical for the long term.

An option that should also be considered is that of going to a builder that will build a basic box that will encapsulate the processing and functional lines. In many cases, the wall panels are prefabricated to fit into a steel I-beam structure. The whole plant can be erected in a very short time.

The requirements of the FDA and the USDA will be satisfied if the building provides

1. Adequate space for equipment installation
2. Adequate space for storage of materials
3. Adequate lighting
4. Adequate ventilation
5. Protection against pests

Fortunately, there is a planning team that will consider all the aspects before finalizing the construction plan. It may be useful for everyone that could be part of such a team to have some rudimentary knowledge of building systems.

13.6.1 Foundations

Foundations are designed to spread the load that needs to be supported over the load-carrying capacity of the earth. The foundations are normally just reinforced concrete pedestals or walls that are poured into forms or trenches up to the finished grade line. If the carrying capacity of the soil is inadequate, piles are driven into the soil to support the foundations.

13.6.2 Supporting Structure

The foundations are backfilled and a supporting structure is erected on the foundations. The supporting structure may be

1. Reinforced concrete columns and beams
2. Prestressed concrete tees on reinforced concrete columns and beams
3. Steel joists on reinforced concrete columns and beams
4. Steel joists on steel beams and columns
5. Poured concrete floors and frame walls

The selection of the structural material will depend on the following conditions:

1. Load-bearing capacity required and the number of stories
2. Size flexibility options
3. Desirable clear spans
4. Cost at the specific location

Using reinforced concrete construction can eliminate ledges that collect dust and require cleaning. The forms must be very well prepared to give a surface that can be left bare or painted after the forms are removed. Any ridge or air bubble hollow in the concrete will add cleaning problems. Ridges should be removed and hollows filled in. These defects should be removed before the concrete has cured.

Steel framing is constructed more easily and quickly and it leaves a structure that is easily modified at a later stage. One of the problems with steel structure is that supporting beams could be located too close to a wall, leaving a space that is too small to clean yet large enough to harbor many problems. If this situation exists, fill the space with polyurethane foam and seal the surface of the foam. Remember the golden rule: Avoiding hollow bodies in a plant will avoid many problems! The use of I-beams and H-beams to support the roof leaves large ledges that can accumulate dust. In bakeries, this "dust" is frequently flour and it can cause severe fungal problems in high-humidity areas. The beams can be sealed by tack-welding covering plates over the area. Square tubing can be used to eliminate the dust-collecting surfaces. The tubing must be sealed at either end and any holes drilled into the tubing must also be sealed.

13.6.3 Walls

Walls should be water, insect, and rodent proof. Rodents are particularly troublesome since they require such small holes to gain entry. A rat needs a hole of about 12.5 mm (½ in.) diameter and a mouse requires a 6.3 mm (¼ in.) diameter hole to gain entry. Rats and mice can climb walls, can fall 20 m without being killed, can climb on wires, and do things unimaginable to gain access to food.

Walls are normally built of concrete blocks and/or brick masonry for the general building envelope. If concrete block alone is used, it should be treated to make it waterproof. The wall finish and additional insulation depend on exposure. A vapor barrier must be incorporated in an insulated wall or damage to the insulation will result.

Ceramic tile of the correct grade that has been applied properly is a cost-effective, hard-wearing surface for most sanitary work areas. It requires careful installation so that there are no hollow openings behind the tiles. Many other durable surface coverings such as fiberglass sheets are used where economy is required. Another advantage is that the panels are 1.2 m wide and there is only one seam every 1.2 m. The panels are hard wearing and can be replaced. The wall surface should be protected against damage from fork trucks, trolleys, and carts. A curb or guardrail will be sufficient.

Wall angles, corners, and junctions of walls and floor should be sealed and rounded to facilitate cleaning. The concave coving is normally from 25 to 150 mm radius.

13.6.4 Floors

Floors are normally constructed over a concrete slab. The poured reinforced base thickness will vary depending upon the load it should carry. The finish of floors varies with the location where it must be used. Whatever the finish, floors should have the following characteristics:

■ Be impervious to the spillages of product, cleaning solutions, hot and cold water, disinfectants, pesticides, and lubricants.
■ Be resistant to damage from fork truck wheels and scuffing from general maintenance equipment and tools. They should also not be damaged by cleaning solutions, hot and cold water, disinfectants, pesticides, and lubricants.
■ Be smooth for easy cleaning and disinfection, but not slippery.
■ Be repairable in sections.
■ Look good and be hygienic.

13.6.4.1 Slope of Floors

Floors should be sloped to avoid accumulation of puddles of water in any area. Floors are normally sloped about one vertical unit per forty or sixty horizontal units.* Common sense will dictate how frequently floor drains should be installed to allow the slope to be small enough that wheeled equipment does not run away. As a rule of thumb, the drains should be about 3 m from the high point of a floor. This means that the ideal would be one drain per 36 m². It is also good to remember that any equipment that needs to be horizontal on a sloped floor should be fitted with leg jacks or adjusters.

The speed with which water will drain across a floor depends upon the roughness of the floor. If the floor is in a multistory building, it is important that an impermeable membrane is used so that water does not drip through the concrete.

Processing floors should be located above ground level to prevent flooding and to reduce drainage problems.

13.6.4.2 Kinds of Floors

13.6.4.2.1 Concrete

Plain concrete is suitable for inside warehouses, storage areas, and some processing facilities. Acids from food or acids generated by microorganisms will etch the surface. High-acid foods such as pineapple, apricots, and tomatoes can cause severe damage to concrete surfaces.

13.6.4.2.2 Metal Plate Floor

These floors are used for extreme service such as shipping areas and loading bays. The 30 × 30 mm plates with a raised pattern to make it less slippery are normally bonded to a concrete subfloor. The epoxy must be formulated to allow movement caused by expansion of the steel tiles. The plates are normally butted when they are installed to minimize the exposure of the bonding agent.

---

* This means about 1 m slope over the length of a football field.

13.6.4.2.3 Monolithic Floor Surfaces

Surface preparation of the concrete subfloor is an important part of the success of a chemical-resistant floor. Sandblasting is an excellent way to prepare the surface. If this is not possible, an acid wash could be used. Two types of coverings are used: monolithic surfacing and paving or tiles.

Monolithic surfacing is a layer of synthetic resin compound that is bonded to the concrete. The four types of resin that are used include epoxy, furan, phenolic, and polyester. Each of these resins will have different protective properties under different conditions and selection is made according to the area where it will be applied and product that will be processed.

Monolithic surfaces will slowly absorb liquids. If the liquid is acid, the acid will eventually reach the concrete subfloor. These coatings should therefore not be used where high-acid products are processed. These surfaces are also damaged by heat, and welding splashes will damage the surface to an unacceptable level. Monolithic floor surfaces are best used in areas that are dry or areas that are wet for only short periods of time. They can also be used where the reactions between the covering and the product are limited.

Monolithic surfaces can be used to

1. Cover floors where impact and thermal exposure are moderate
2. Cover places where the installation of tiles is impossible
3. Cover floors of processing areas with a limited life span
4. Reduce initial cost
5. Ensure nonskid surfaces

It is important that a waterproof membrane is installed between the structural slab and the fill or leveling slabs. It is also very important that a proper drainage system be provided.

13.6.4.2.4 Tile or Paver Floors

Tile or paver floors are used in the food industry because they are very durable, attractive, and versatile. The success of the floor is dependent upon selection of the proper thickness of tile, grout, bedding system, and installation method.

Two kinds of tile are used. Quarry tiles are cheaper than ceramic tiles and offer good impact resistance and very good resistance to chemical attack. Ceramic tiles have the same properties with an attractive finish. Tiles are 12.7 or 19.0 mm thick and should only be used in areas subject to light traffic.

Pavers or bricks are thicker than 38 mm and are suitable for general use in most areas of a food plant. Thicker brick or tile should be used where loads are excessive or where it is necessary to dissipate heat.

The selection of the type of grout that is used should be based on such service conditions as temperature, exposure to process materials, maintenance practices, and mechanical requirements.

13.6.4.3 Drains

Floor drains in the processing area allow for continuous cleaning and will thereby contribute to the overall sanitation of the operation. It is important that the drains be correctly spaced and located. The floor drain is designed to remove wastewater from the processing area. The wastewater will frequently contain large amounts of food particles that will increase the biological oxygen demand (BOD) if washed into the drain. It is important to remove solid waste before it reaches the drain. Cleaning by hosing into a drain comes very naturally to most people and this habit should be discouraged with cleaning staff.

Drains and the drainage system should be capable of removing the maximum load that can be generated. This is normally during the cleaning operation. This capacity will avoid drain overload and contamination of working areas. Sanitary drains containing fecal matter should not be routed through food processing or storage areas. Wastewater drains and sanitary drains should also not be connected on site. In some cases, the treatment plant will keep sanitary waste and industrial waste separate.

In some cases, floor drains are placed on the periphery of the room in a drainage trench. In other cases, drains are placed under or near processing machines that generate a large amount of wastewater. With cleaning-in-place systems, the amount of water to be drained from machines has been reduced a lot. In many modern plants, floors are essentially dry except during cleaning.

13.6.4.3.1 Floor Drains

Floor drains are normally set in the floor in a way that the grating is about 3 mm below floor level. The capacity of floor drains should be large enough to handle present and future needs. There are many types of drains available and selection of a specific drain pattern will be dictated by requirements and cost.

13.6.4.3.2 Gutter Drains

Gutter drains have smooth vertical walls with a round or coved bottom. The minimum slope of the channel should be 1 in 100. The normal size for the channel drain is about 150 to 300 mm wide and 150 to 200 mm deep. Deeper gutters are difficult to clean.

The gratings of gutter drains are normally set slightly above floor level. The grating must be strong enough to handle heavy traffic like fork trucks. The gratings must be in short sections since they need to be lifted out for cleaning. The openings of the gratings flare outward toward the bottom to ensure that any food pieces that can enter the grating will not get stuck in it.

13.6.5 Windows

Windows in the processing area should have glass that is unbreakable, such as laminated, reinforced, or polycarbonate sheet. Windows that can open should be fitted with screens to avoid entrance of birds and insects. Outside windowsills should allow water to drain away from the window. Inside sills should slope between 20° and 45° to facilitate cleaning. In high-risk areas, windows should not be installed. Open windows with screens can still be a source of airborne bacterial contamination.

13.6.6 Doors

Doors are placed in buildings to allow access to people and equipment but keep contaminants, rodents, and insects out. They are also a barrier to maintain the controlled environment inside the building. Doors need to fit inside the jambs with limited clearances. Doorjambs should be metal and, if hollow spaces exist, the spaces should be filled with mortar or foam.

The doors should be steel that is continuously welded to seal the inside cavity. Solid-core doors should be clad with stainless steel. The problem with them is the vulnerability of wood to insect and rodent damage. In some processing areas, I have seen solid aluminum doors. They are, however, prone to buckling after some use. Plastic doors are becoming a favorite in processing areas.

Doors that open to the outside are a particular problem. It is always bad practice to design direct access to and from the outside for a processing area with everyday use. Such an exit should be used in emergency situations. Allowing people to enter the processing area without proper sanitation of boots and hands can seriously compromise hygiene. Meat packing plants where animals are slaughtered are an example of some very "dirty" areas in the holding pens adjacent to very "clean" areas in the deboning rooms. Allowing direct access from the one area to the other should be avoided.

It is a good idea to locate an area light at least 15 m away from the door. This will attract insects to an area 15 m away from the door. The practice of lighting a doorway attracts insects and can allow a whole lot of them to enter the facility.

13.6.7 Piping

The position of water, compressed air, refrigeration, and steam lines should be carefully considered. It is possible to install trapeze hangers to route any number of lines through the plant, delivering services to most of the equipment. Where equipment is not in direct reach of the main route, a short branch will not cost too much. Additional outlets and tees can be provided for future machinery.

Using this main route for most of the utility lines in the plant allows for easier inspection and maintenance. Having the lines in a simple and, for the most part, straight configuration also makes it easier to insulate the refrigeration and steam pipes. Wherever the lines cross the processing line where food is exposed, a stainless steel protective roof should be supplied under the service line trapeze.

In some cases, the services are run on the outside of the processing area walls. This allows for servicing of the main lines without entry into the processing area, thereby avoiding a source of contamination. The layout inside the processing area will require many lines running from the main line to each piece of equipment that needs to be serviced. In the case of steam, it is good to remember that each steam line will have an associated condensate return line. Any additional line therefore becomes a double addition.

13.6.8 Electrical

The electrical system in a processing plant should be as flexible as possible. Motor control centers, situated throughout the plant, will control the power distribution to the processing motors in the plant. If the system was planned with a number of spare sections, new motors can be added very easily. All that needs to be done is the installation of flexible conduit to the new motor and running the wire.

13.6.9 Ventilation

Ventilation is the supply of fresh, conditioned air to replace unwanted air. Conditioning can include alteration of moisture content, change of temperature, and filtering to remove particulates and organisms.

Within the processing area, ventilation will remove obnoxious odors, moisture, and heat and replace them with air that is free from contaminants and will increase the comfort level of the workers. The amount of air is calculated as a replacement volume. Depending upon the production processes, the air can be replaced from six to twenty times per hour. It is also advisable to keep the processing area under a slight positive pressure. This will ensure that processing area air flows out when a door is opened.

Special air is required in areas where baby formula is handled or where aseptic operations take place. In this case, the air will be filtered through special filters that will remove organisms. The processing area must be under positive pressure at all times so that no organisms can enter from adjacent processing areas.

13.6.10 Hand-Cleaning Stations

In many plants, the installation of hand-washing stations at every entrance into the processing area has resulted in much improved hygienic standards. In high-risk production areas, the workers have to remove their aprons when they leave the processing area and they will be issued new aprons when they enter it again. They will have to wash their boots and then their hands before they re-enter the processing area. The entrance will have a boot bath filled with sanitizing solution to prevent unwanted organisms from being carried into the processing area on boots. Modern hand-washing stations utilize fine jets of water to clean hands.

13.7 Facilities Layout

A plot plan will show how the buildings, parking lots, and driveways fit on the lot. It will also show highways, utilities, drains, electrical systems, and any other information that is important to the project. City and county building codes might require frontage roads and have certain boundary restrictions that must be displayed on the plan.

Start with the plot plan and add the main roads that border the property. Determine where the access roads will enter the property. Indicate the utilities on the plan. Place the building so that the front faces a road. Expansion will occur behind the building. Indicate where receiving and shipping will be and connect this area to the main road. Draw the employee parking and designate its access from the road. Provide parking for visitors and management.

13.8 Summary

Planning a new processing facility is a long-term project that requires projections of many years into the future. It frequently happens that between the time of planning and the time of moving into the facility, the "new" technology that was planned for is totally redundant. Planning a flexible facility will allow for making continuous changes regarding the machines that will finally be installed. By using a checklist it is possible to make some calculated projections.

1. Establish the amount of product that will be processed in one 40-hour week.
2. Establish the transportation requirements for the raw material and the finished product.
3. Estimate the size of the facility and service buildings and decide on a lot size. Multiplying it by two should give the real required size for present and future needs.
4. Contract for a geological survey of the plot and determine where sewer, water, gas, and power mains are located.
5. Plan the building with detailed layout of equipment, ergonomic studies, facilities, warehouses, and other facilities.
6. Ensure an economical and efficient materials handling system.

7. Review the load-bearing capacity of the structure when the size of storage tanks and heavy equipment is taken into account.[*]
8. Determine the surface finishes that will limit maintenance costs.
9. Obtain approval from all authorities with jurisdiction.
10. Review budget estimate and the effect on the profit and loss statement.
11. Determine the total requirements for steam, compressed air, water, chilled water, gas, electricity, lighting, drains, and refrigeration.
12. Determine ventilation for requirements for comfort.
13. Determine sanitary and personnel requirements, including toilets, showers, lunchrooms, etc.
14. Determine the maximum number of workers per shift.
15. Review the qualifications of consulting architects and previous engineering experience.
16. Review qualifications of the contractor and construction team.
17. Prepare financial feasibility schedule and review with financial sources.
18. Review semifinished design with department heads.
19. Review office needs and computer needs.
20. Review materials handling and product piping or storage design.
21. Review central cleaning system.
22. Review fire prevention and detection equipment.

---

[*] Many high-rise apartment buildings had problems when the waterbed was first invented and installed in all the apartments, one on top of the other. Calculate the extra load on the supporting columns.

Chapter 14

Environmental Issues

14.1 Costs and Benefits of Environmental Compliance

The sale of products or services is based on the four Ps of marketing—product, price, place, and promotion. Internally, consumer satisfaction used to be assured by production and quality management, where we relied on the few simple steps that can be summarized as who, what, where, when, and how. By following these simple rules, the company could satisfy the needs of most consumers.

Corporate social responsibility used to be considered as corporate philanthropy, moral obligation, or pure altruism. This point of view changed when people realized that social responsibility is a source of competitive advantage. What used to be money given to a good cause became a significant corporate resource regarding competitive relationships (Piercy and Lane, 2009). This new approach is in stark contrast with the point of view of the neoclassical economists, who defined the social responsibility of businesses to maximize the profits of the owners (Balabanis, Phillips, and Lyall, 1998).

14.2 Mandated Environmental Costs

The developed countries are formulating and extending environmental policies. In the United States, the environmental policies are also under continuous scrutiny to ensure that the country will keep up with the Scandinavian countries that are leading the way. The likelihood of a mandatory and perhaps a punitive greenhouse gas mitigation program is a strong possibility for the United States. The cost of this policy is difficult to determine and the distribution of the winners and losers from implementation of this policy is impossible to predict since its actual total cost will depend upon its final design. Such a policy will have a bigger and wider impact on companies and consumers than any of the existing environmental policies (Aldy and Pizer, 2009).

The foundation for future climate policy was provided in the 110th Congress in 2007–2008 where members of the US Senate identified their principles for a cap-and-trade program (Aldy and Pizer, 2009). According to the information on the Environmental Protection Agency (EPA) website (EPA, 2010), the cap-and-trade environmental policy will deliver results when a mandatory cap is placed on emissions. There will be flexibility regarding the sources of compliance and a rewards program for innovation and early action. The program will also provide strict environmental accountability but it will not inhibit the economic growth rate.

14.2.1 Change

The rising cost of environmental compliance is a function of the company's impact on the environment. It is also a symptom of the inadequacies of the company's efforts to optimize resources by minimizing waste. Remember, Lean production is essentially green production since the output stays the same while the inputs decrease. Environmental compliance costs can be ameliorated by reducing the waste streams and formulating a Lean strategy for the production facility.

Change from the current strategy is a movement away from what is toward a future state of what could be. This organizational change causes enhanced organizational performance that can be measured in terms of increased profitability, greater responsiveness to the market, and improved corporate sustainability. The greatest value of organizational change is embedded in the ability to alter the organization's identity regarding ecological and operational strategies as a method to enhance performance (Vithessonthi, 2009).

14.3 Corporate Social Responsibility

Corporate social responsibility embraces multiple stakeholders within corporate obligations. The stakeholders include stockholders, workers, the community, and the environment. The outcomes of the corporate mission with regard to social responsibility target economic, legal, ethical, and philanthropic dimensions of doing business (Dawkins and Ngunjiri 2008). We live in an era that is characterized by a significant emphasis on ethical norms and the stark reality of unscrupulous and unethical business practices.

A consumer's selection process is based on many product qualities, including an assessment of the manufacturer's social, organizational, behavioral, and technological choices. Human values and social responsibility are gaining paramount importance to ensure continued support by consumers. Management has to balance these complex relationships to ensure continued growth and development of the company while it also focuses on the protection of the environment at national and international levels (Comite, 2009). Because human values are becoming more relevant and consumers focus on ethical factors, management must develop policies that benefit local stakeholders, employees, the local environment, and the image of the company on national and international levels. The concept of ethics is multi-form, being both abstract and discordant at the same time. It behooves companies to adopt an increasing number of environmental strategies to react to legislative norms, on the one hand, and to the growing market trends that favor the products of companies that are more sensitive to environmental problems on the other (Comite, 2009).

Corporate social responsibility is no longer an option since the classical economic paradigm will no longer sustain continued growth of the business. The divergence between a company and its socially responsible behavior is becoming vague since marketplace performance is concomitant upon consumers' impressions of the company's responsible behavior (Bucholz, 1991).

Bucholz (1991) lists five key elements for corporate responsibility:

- Corporate responsibility goes beyond the production of goods and services for profit.
- A corporation is responsible to solve social problems, especially if it created them.
- A corporation has a broader constituency than stockholders.
- A corporation has impacts that go beyond simple marketplace transactions.
- A corporation serves a wider range of human values than what can be captured by a simplistic focus on economic values.

Top management is ultimately responsible to influence the formulation of an ecologically friendly sustainable strategy (Vithessonthi, 2009). Sustainable development embraces the ideals of economic growth and the protection of the environment. This consciousness of sustainable development permeates the consciousness of governmental agencies, international policy makers, multinational corporations, and consumers within their respective societal boundaries. Society knows that development of production systems that meet the needs of present consumers without compromising the ability of future generations to meet their own needs is the only way to sustain the future of everybody. It is important to realize that both economic and ecological factors will manipulate the company's future ability to do business. There has thus been an irreversible paradigm shift from pure economics to sustainability (Beckmann and Pies, 2008).

14.4 Legislative Issues

Environmental compliance is no longer an option for companies that want to ensure their long-term survival. Legislators made compliance and environmental stewardship a part of their promise to the voters. The United States has also entered into agreements with other world leaders to become more proactive regarding climate change.* This attention to environmental impact caused by US corporations will make it difficult and even impossible for a group of lobbyists† to influence environmental policies. It must also be emphasized that environmental regulations are part of the reason why corporations moved their operations to other countries.

Environmental noncompliance can also be an extremely dangerous path for a US company. The US public supports the reduction of emissions and they are the consumers of the products. In cases where a company's products are not directly utilized by consumers, the consumers may place pressure on the companies that use that product as a component, to use only components from companies that are EPA compliant. As an example, when components are manufactured that Ford uses in its automobiles, it would be suicide if the company were to be alienated because of an environmental issue since it would lead to reduced production with concomitant reduction in staffing. In the present economy with the large number of unemployed people, adding to the unemployment because of noncompliance with environmental legislation will alienate the politicians that may help the company to compete successfully for federal contracts.

Compliance with all the regulations will obviously necessitate the upgrade of some production facilities. Since upgrades are expensive, it may be more prudent to use legislators to earmark some form of financial grant or stimulus for the company to become more compliant. In this scenario, the company can start advertising its good citizenship by complying with all the regulations for the welfare of the people living close to the plant. This public emphasis of corporate social responsibility will also enable a more competitive approach to the companies that use the raw materials or components from compliant producers. The changes required to the plant will also require a long-term construction project that will enable the creation of employment for many people from the area. This could alleviate the social dilemma for people that are already employed by the company.

All this will be seen as social and ethical responsibility of the company, and this should drive the stock price higher, thereby ultimately benefiting the stockholders. It is a win–win situation for everybody involved.

---

* The correlation between human activity and climate change is a contentious issue. Whenever you are unsure about this issue, keep in mind that we know that good stewardship of our resources and sensitivity to environmental issues are the best possible way to ensure our continued well-being.

† Lobbying is a system where a lobbyist will use campaign contributions to coerce legislators to support legislation that will benefit the corporation that hires the lobbyist.

14.5 Environmental Compliance of Operations

14.5.1 Introduction

Compliance with EPA regulations is of prime importance for any company that wants to be responsive to environmental issues. Compliance with the EPA regulations should be seen as the first step in environmental responsibility. We live in an era that is characterized by a significant emphasis on ethical norms and the stark reality of unscrupulous and unethical business practices. This scenario, coupled with overwhelming information distribution, does not allow any business the luxury of doing business incognito. Whatever happens in a company will sooner or later be made public. As such, air and water are not environmental issues any longer; they are business issues. The production history of a company and its effects on the environment are inextricably linked and successful companies use this to their benefit (Manning, 2004).

Emissions control is of prime importance and emissions can be classified according to diverse criteria according to the nature of the emissions. Some emissions are fugitive, like sulfur dioxide and carbon dioxide, while others are persistent, like an oil spill. This classification will determine how long the emissive pollutant lasts in a particular place. Another aspect regarding emissions is the difficulty to determine the exact source of an emission. Point source emissions are observable to an auditor, while pesticides in water can be caused by runoff from farm aerial applications or from a company fumigating a raw material (Macho-Stadler, 2007).

Many companies in the United States are participating in voluntary government-sponsored pollution reduction programs. The programs, sponsored by environmental agencies, aim to encourage industries to take a holistic view of their impact on the environment. Many companies are taking proactive steps to improve their management of the environment by utilizing management systems that balance operational decisions with environmental impact information (Sam, Khanna, and Innes, 2009).

In 1991 the US EPA sponsored the 33/50 voluntary pollution reduction program. The goal of this program was to reduce emissions of seventeen high-priority toxic chemicals by 33 percent by the end of 1992 and 50 percent by the end of 1995 based on the levels of 1988. The approach was to prevent pollution rather than abate pollution that had already occurred. Total quality management is a tool that exemplifies this ethic since it emphasizes prevention over detection and continuous improvement by minimizing defects and waste (Sam et al., 2009).

14.5.2 Organization and Staffing for Environmental Compliance

The key to success in environmental issues is embedded in the people. It is essential that top corporate management is involved in and committed to the environmental management program. The organization also needs to be structured in a way that will allow successful implementation of the environmental management program. The environmental function must be integrated in both the risk engineering and safety functions of the organization (Friedman, 2006).

"Companies lose money because they treat pollution control and plant operations as separate concerns. It costs less in the long run to make environmental and plant managers true partners in finding compliance solutions" (Singh, 2000, 91). There has been a growing shift in environmental thinking. What used to be compliance to regulations became an environmental corporate stradegy as part of strategic management. This affects decisions regarding product development, future process technology, and total quality programs. This paradigm shift leads to continuous adjustment, adaptation, and innovation. Environmental management is no longer a side issue, but rather the very essence in the way companies do business (Dechant and Altman, 1994).

14.5.3 Strategic Management

Successful environmental management depends on a strategic shift in the way that a corporation conducts business. The corporations that operate within the paradigm where assumptions, values, and beliefs are focused on profitability and economic growth at the cost of ecological concerns need to redefine their strategies. The alternative to the technocratic understanding of a corporation's relationship to nature will place the environment at a position of being intrinsically valuable and inseparable from the long-term profit-seeking strategy (Starkey and Crane, 2003). The strategies for sustainability should at the very least comply with the law and ensure that that each employee is able to report a violation of the policy in a way that will ensure no direct retribution (Friedman, 2006).

Top managers play a dominant role when corporate strategy is formulated. The change in the corporation's ecological performance will modify the corporation's identity, structure, operations, and human resource policies. The change will move the organization from the present toward a future state that is environmentally responsible. This organizational change will cause enhanced organizational performance that can be measured in terms of increased profitability, greater responsiveness to the market, and improved corporate sustainability. The greatest value of organizational change is embedded in the ability to alter the organization's identity and ecological and operational strategies as a method to enhance performance (Vithessonthi, 2009).

14.5.4 Organizational Structure

Organizations are frequently structured as centralized or decentralized entities. The decentralized structure with a small corporate staff works well in most cases. In both the centralized and decentralized structures, the corporate environmental protection staff must provide the leadership to ensure that the environmental policy permeates to all levels of the organization. To facilitate environmental management, a structure must be developed where all employees take responsibility for the environmental impact of the organization (Friedman, 2006).

Each divisional environmental department is charged to define the areas of responsibility for its specific environmental problems. The people in the divisional environmental departments identify programs and issues and recommend corrective actions and strategies that will ameliorate the problems (Friedman 2006). In the case of packaging, the ISO 14000 standards include problems with the packaging materials in countries of destination, so identification of environmental problems has to include problems across the supply chain.

Forward-looking companies are implementing structures based on different types of teamwork, including total quality management (TQM), employee involvement (EI), or self-management work teams (SMTs), within the hierarchy to ensure that all employees are involved in the management program. The corporate organizational structure will delineate the hierarchy and reporting relationships between different functions and levels in the organization (Waters and Beruvides 2009). The duties for individuals at each level of responsibility and functional area are summarized in a detailed organization chart (Friedman, 2006).

To transfer the corporate environmental mission to the departmental environmental teams, it is important that the environmental goals and objectives are clearly defined. People should also be given specific responsibilities pertaining to the areas in which they work. The establishment of reporting relationships and communications channels is of great importance for success of the efforts. It is also important that resources such as time, equipment, budget, and expert people are made available at the different levels of the organization.

14.5.5 Environmental Audits and Review

Environmental audits are intended to aid in the identification of problems. The audit is an objective, orderly, periodic, and fully documented review of operational practices that may impact the environment. The environmental audit is similar to a financial audit but with an obviously different purpose (Friedman 2006). The environmental audit report addresses the present environmental performance, the corporate environmental vision, and the problems that must be looked at in the future.

14.5.5.1 Solid Waste

For solid-waste streams that are nontoxic, the EPA Auditing Protocols Subtitle D will be used as a guideline. The first step in the audit will be an identification of solid-waste streams and an assurance that the solid waste is nontoxic. The solid-waste streams and materials are then separated into materials that can be recycled and the materials that will be moved to a landfill. All materials will be stored in such a way that they do not cause a fire, safety, or health hazard. The removal and collection of the material will be done in a way that will ensure the health and well-being of the personnel operating the system. The equipment that is used for movement and collection of the material will have a suitable cover to prevent spillage. The equipment will be adequately maintained to ensure no unmanaged spillage of waste (EPA, 2000).

Solid waste that is deemed hazardous will be transported off the site by the company that handles those materials. The hazardous waste will be packaged in accordance with the applicable Department of Transportation regulations on packaging under 49 CFR parts 173, 178, and 179 prior to being loaded on the transport vehicles. Each package of hazardous waste that is placed on the transport will be labeled according to Department of Transportation regulations on hazardous materials under 49 CFR part 172 (EPA, 2000).

14.5.5.2 Recycling

If the company does not employ more than 100 office workers, paper separation does not have to be done to comply with EPA requirements. The company will, however, be involved in the local recycling programs to reduce the volume of solid-waste materials going to the landfill. The high-grade paper will be recovered and placed in separate recycling bins. If the company generates more than 10 tons of corrugated containers per month, the corrugated paper will also be held separate for recycling purposes according to the EPA regulations (EPA, 2000).

14.5.5.3 Air

The pollutants in air that have the greatest effects on the health and safety of people in the areas around the processing plants include ground-level ozone, carbon monoxide, sulfur oxides, nitrogen oxides, and lead (EPA, 2008). If the company produces processing steam by burning coal, all of these pollutants can be present in the smokestack. The pollutants that occur in the smoke emissions are within the Air Permit Program administered by EPA (EPA, 2009).

14.5.5.4 Water

In a food factory, the four impurities in water that causes public health concerns are organic chemicals, inorganic chemicals, turbidity, and microorganisms. Wash and rinse water contain organic contaminants that include insecticides, herbicides, fungicides, and soaps. All these substances can be harmful to the ecology if they are present above certain concentrations in effluent streams. Any substance or organism that could pose a threat to the ecosystem will be rendered harmless before the wastewater is introduced into the surface water. Suspended solids and particulates will be removed. Soluble solids will be diluted to a level where they will not harm the ecosystem. Nitrates and phosphates will be removed to stop excessive proliferation of fungi in the water. The national pretreatment standards for water effluents will be used as guidelines to limit quantities or concentrations of pollutants, which may be discharged to the local water treatment plant (EPA, 2006).

14.6 ISO 14000 Series Environmental Management Standards

In March 1987, the Brundtland Commission of the World Commission of the Environment as part of the nongovernmental organizations (NGOs) associated with the United Nations proclaimed: "Humanity has the ability to make development sustainable to ensure that it meets the needs of the present without compromising the ability of future generations to meet their own needs" (UN 1987, 16). This became the mission of the World Commission on the Environment and Development (Stenzel, 2000).

The International Organization for Standardization (ISO) adopted the attainment of sustainable development as a major goal in the ISO 14000 Series Environmental Management Standards in 1996 (Stenzel, 2000). The regulations are contained within the ISO 14000 family of regulations. The first two standards, ISO 14001:2004 and ISO 14004:2004, regulate environmental management systems. The other ISO 14000 standards and guidelines address specific environmental aspects such as labeling, performance evaluation, life cycle analysis, communication, and auditing (ISO, 2010).

The international acceptance of the ISO 14000 standards assures that companies pursuing unrestricted international trade are complying with the standards accepted by ISO. Business managers and their legal and economic advisors view ISO 14000 as a consumer-driven mandate to environmental protection. In this sense, it is different from the regulations promulgated by the EPA that use a command and control type of regulation. The author concludes that the nonregulatory ISO 14000 regulations can provide an incentive framework for companies to advertise their environmental concerns while supporting the goals of the EPA (Stenzel, 2000).

The ISO 14000 family of regulations is based upon the principles drafted by the Coalition for Environmentally Responsible Economies (CERES). The environmentalists wanted business to include environmental principles in their environmental management systems (Stenzel 2000). The principles are stated in Table 14.1.The Technical Committee of ISO took the CERES recommendation under review but excluded the final principle because most businesses deemed that too harsh. The ISO 14000 standards act as a bridge program among different environmental management and auditing programs that are used globally. Many of the government-sponsored and industry-based certification programs use different standards and

Table 14.1 CERES Environmental Principles

	Environmental Issue
1	Eliminate or minimize harm to the biosphere
2	Use of sustainable resources
3	Minimize waste and recycle when possible
4	Use renewable energy resources and maximize energy efficiency
5	Minimize risks to employees and the communities where the business operates
6	Market safe products and services and inform consumers of the environmental impacts of the products and services
7	Address injuries and, when damages occur, restore the environment to a feasible extent
8	Disclose to all concerned any information about incidents posing environmental or health and safety harm
9	Ensure that at least one member of the company's board of directors is a person with demonstrated environmental commitment
10	Conduct and publicize an annual self-audit regarding implementation of the ten principles and to assess compliance with all applicable laws and regulations worldwide

processes. This causes an increase in costs and makes compliance difficult for companies operating in multiple countries (Stenzel, 2000).

To be registered as an ISO 14000 compliant company, the company must create an environmental management system. The company must then audit itself to demonstrate that it complies with the environmental statutes and regulations of the countries in which it operates as a producer and vendor. Finally, the company must demonstrate its commitment to prevent pollution and continuously improve its environmental protection acts (Stenzel, 2000).

The environmental management system is a set of procedures to review compliance with the applicable environmental laws as well as methods to measure the company's own procedures for identification and resolution of environmental problems. The system must also address the issue of how the company's workforce will be engaged in a commitment to improved environmental performance. For ISO 14001 compliance, the company's environmental management system must include a summary of the federal, state, and local legal standards and laws that the company must comply with (Stenzel, 2000).

According to Stenzel (2000), ISO 14001 certification causes many rewards for the company that implements it. The most important result that companies hope for is more lenient filing and monitoring requirements by the EPA. In the case of violations of an environmental law, the company also hopes that it will receive less severe penalties. An additional benefit accrues because of the company's enhanced relationship with the EPA.

Many consumers also choose to do business with companies that demonstrate their environmental stewardship and accountability. Investors may also choose to do business with environmentally responsible companies. There is also the possibility that insurance premiums may be reduced for a company that is ISO 14000 compliant. The implementation of the environmental management system will reduce the likelihood of toxic spills that can endanger employees and the local community. The procurement of funding from financial institutions is also easier for ISO 14000 compliant companies. There are benefits from cost savings resulting from waste reduction, use of less toxic chemicals, energy reduction, and recycling.

14.7 Corporate Responsibility

"The economic paradigm will continue its dominance as long as human beings consider themselves to be the center of life on earth" (Buchholz, 1991, 19). Corporations consider social responsibilities only after they have exhausted their marketplace performance. Social responsibility is thus brought into place only when it can augment the marketplace performance. This Band-Aid approach to corporate social responsibility should change since any divergence between the successful operation of a business and its social responsibility will lead to a decrease in performance (Buchholz, 1991).

The historical view of business was as a provider of goods or services to consumers at affordable prices. To be successful, the business must economize resource utilization to maximize profits. The business also provides jobs and income for employees and thus benefits society by contributing to the wealth of the society. In this way the social performance of the business is linked to the performance in the marketplace. Thriving business performance in the marketplace and agreeable corporate social behavior are intertwined (Buchholz, 1991).

The old paradigm is no longer acceptable and businesses are required to clean up pollution, to provide safer work environments, to produce products that are safe to use, to promote equal opportunity, and to promote human welfare in the society that they serve. Many businesses look at their profit margins and keep on disposing of waste as cheaply as possible, thereby causing serious pollution. Businesses frequently hire the most qualified people for the job, and this action perpetuates discrimination against minorities and women in the communities in which they are situated. Society is starting to question the social responsibility of corporations. In the past two decades, the points of divergence between what is good for profitability and what society expects of the businesses became obvious. Advocates for social responsibility made a case that corporations needed to take the social impacts of the business into account for long-term strategic planning. This allowed businesses to respond to the social expectations of the community (Buchholz 1991).

Society has changed its perspectives about nature and the changes are reflected in the policies that have been adopted as a nation. Even during the pioneering days when natural resources were exploited on a great scale, the conservation of the natural heritage was recognized by policy makers and the first of the national parks was established. The conservation movement attempted to curb the reckless exploitation of natural resources. In later years, this was extended to the protection of the ozone layer. It became apparent that environmental problems are interrelated. Environmental problems such as global warming and ozone depletion threaten humanity globally. Although social responsibility is normally regional, and that is fine, the larger scale threats to the ecology require more global control. "It is difficult to talk about air pollution in one country and efforts being made to reduce air pollution without talking about other countries' problems" (Buchholz 1991, 28). This is also true for water pollution and waste disposal problems. Space travel allowed us to view the Earth and the effect that we have on parts of the world. It also allowed us to realize that life on Earth is not replicated on other planets. We have a unique environment that may be more fragile than we previously thought. The Earth has finite resources and people have a responsibility to manage the human exploitation of it (Buchholz 1991).

It is not possible for economic growth to occur without the adequate environmental conditions to support that growth. There is no possibility to make trade-offs between economic growth and environmental protection. The environment must be protected and maintained to allow continued economic growth. Without environmental husbandry, all that will be left is a ravaged Earth that may not be able to sustain life. Any economic growth that subverts the conditions for that growth is not sustainable (Buchholz 1991).

14.8 Application of the Principles

The continued well-being of any organization depends upon continued care of the environment. Consumers will no longer do business with companies that threaten the planet. It does not matter if the threat is real or perceived; if consumers deem a corporation as socially irresponsible they may stop buying that company's products or services. Marketing departments know this and exploit the good environmental stewardship of one company against the bad stewardship of an opposition company. To be successful in the marketplace, companies are forced to be more environmentally friendly.

To be environmentally responsible requires planning. It is not economically feasible to separate manufacturing operations and planning from environmental planning. A holistic approach to integrated operations and environmental planning gives the highest return on investment. To integrate environmental systems with the operational systems, companies should review their operations and make appropriate adjustments. They should also find the optimal time to install pollution control equipment. This is frequently at the time when the plant is upgraded or when the plant is expanded. The production technology must also be upgraded at the same time when the operational systems are expanded or changed. Ensure that environmental costs are allocated to the different production systems appropriately. This is the only assurance that the total production and environmental costs are allocated to each individual product. Finally, the business and environmental decisions must be integrated to ensure maximum profits and minimum environmental costs.

Environmental legislation is localized in a country. Whatever legislation the United States may have has no value in any other country. For this reason ISO 14000 has been devised. In the simplest form, this set of regulations concerns the environmental impact of the production system in the country of origin and the environmental impact of the packaging and the product in the country of destination. Companies have the ability to ensure good environmental stewardship of their trading partners by mandating to do business only with ISO 14000 compliant companies. In this way, environmental concerns can become global.

References

Aldy, J. E., and Pizer, W. I. (2009). Issues in designing US climate change policy. *Energy Journal* 30 (3): 179–209.

Balabanis, G., Phillips, H. C., and Lyall, J. (1998). Corporate social responsibility and economic performance in top British companies: Are they linked? *European Business Review* 98 (1): 25.

Beckmann, M., and Pies, I. (2008). Sustainability by corporate citizenship. *Journal of Corporate Citizenship* 31:45–57.

Bucholz, R. A. (1991). Corporate responsibility and the good society: From economics to ecology. *Business Horizons* 34 (4): 19–31.

Comite, U. (2009). The evolution of a modern business from its assets and liabilities statement to its ethical environmental account. *Journal of Management Research* 9 (2): 100–120.

Dawkins, C., and Ngunjiri, F. W. (2008). Corporate social responsibility reporting in South Africa. *Journal of Business Communications* 45 (3): 286–307.

Dechant, K., and Altman, B. (1994). Environmental leadership: From compliance to competitive advantage. *Academy of Management Executive* 8 3): 7–20.

EPA (2000). Protocol for conducting environmental compliance audits of facilities regulated under subtitle D of RCRA. Retrieved on March 17, 2010, from http://www.epa.gov/compliance/resources/policies/incentives/auditing/apcol-rcrad.pdf

———. (2006). General pretreatment regulations for existing and new sources of pollution. Retrieved on March 18, 2010, from http://www.epa.gov/npdes/regulations/streamlining_part403.pdf

———. (2008). Cleaning up commonly found air pollutants. Retrieved on March 18, 2010, from http://www.epa.gov/air/caa/peg/cleanup.html

————. (2009). Air permits program. Retrieved on March 18, 2010, from http://www.epa. gov/air/oaqps/permjmp.html

———— (2010). Cap and trade. Retrieved on February 16, 2010, from http://www.epa.gov/ captrade/index.html

Friedman, F. B. (2006). *Practical guide to environmental management,* 10th ed. Washington, DC: Environmental Law Institute.

ISO. (2010). ISO 14000 Essentials. Retrieved from http://www.iso.org/iso/iso_14000_ essentials

Macho-Stadler, I. (2007). Environmental regulation: Choice of instruments under imperfect compliance. *Spanish Economic Review* 10:1–21.

Manning, D. J. (2004). Benefits of environmental stewardship. *Review of Business* 25 (2): 9–14.

Piercy, N. F., and Lane, N. (2009). Corporate social responsibility: Impacts on strategic marketing and customer value. *Marketing Review* 9 (4): 335–360.

Sam, A., Khanna, M., and Innes, R. (2009). Voluntary pollution reduction programs, environmental management, and environmental performance: An empirical study. *Land Economics* 85 (4): 692–711.

Singh, J. (2000). Making business sense of environmental compliance. *MIT Sloan Management Review,* 41(3), 91–100.

Starkey, K., and Crane, A. (2003). Toward green narrative: Management and the evolutionary epic. *Academy of Management Review* 28 (2): 220–237.

Stenzel, P. L. (2000). Can the ISO 14000 series environmental management standards provide a viable alternative to government regulation? *American Business Law Journal* 37 (2): 237–298.

Vithessonthi, C. (2009). Corporate ecological sustainability strategy decisions: The role of attitude towards sustainable development. *Journal of Organisational Transformation and Social Change* 6 (1): 49–64.

Waters, N. M., and Beruvides, M. G. (2009). An empirical study analyzing traditional work schemes versus work teams. *Engineering Management Journal* 21 (4): 36–43.

UN. (1987). Development and international economic co-operation: Environment. Retrieved from http://www.un-documents.net/our-common-future.pdf

Chapter 15

Safety

We have to recognize that, done right, regulation protects our workers from injury and that when we fail, it can have disastrous consequences. I believe we can bring back common sense and reduce hassle without stripping away safeguards for our children, our workers, our families.

President Bill Clinton, February 21, 1995

15.1 Workplace Safety

Accidents do not happen; they are caused by unsafe conditions or unsafe actions. Someone is always to blame when an "accident" occurs. The lawyers will make sure that the company pays for its lack of removing unsafe conditions. If a person is to blame by unsafe actions and the company suffers enormous loss, the company has to foot the bill. For these reasons, every supervisor and manager must be ever vigilant to make sure that unsafe conditions and actions do not occur.

The Occupational Safety and Health Administration (OSHA) was created in 1970 to help industry identify any unsafe conditions. An unsafe condition is not limited to an open hole through which one can fall; it includes dust or toxin in the air that can cause health problems too.

In the food industry the use of hazard analysis critical control points (HACCPs) is well established where product is concerned. Thinking along the same lines, it could be extended to cover workers (safety) and plant (preventative maintenance).

15.2 Causes of Accidents

The basic premise for safety is that accidents can be avoided. An accident cannot be attributed to an unsafe act or an unsafe condition. It is normally a sequence of events and actions happening within the specific conditions that led to the accident. By anticipating the sequence of events that could lead to an accident, one or more items from the sequence could be removed and the accident could be prevented. This is the basis of industrial safety.

Accidents occur because of improper contact with machinery, lifting and handling of bulk materials, contact with electricity, and contact with hazardous materials. Fortunately most accidents are preventable by rigorous management of staff and maintenance of equipment. A few accidents are inevitable.

There are a number of basic causes for accidents:

1. Supervisor did not give adequate instructions or did inadequate inspections
   a. Failure to give required instructions
   b. Giving wrong instructions
   c. Not appointing someone to oversee safety
   d. Failure to determine safe conditions before the work started
   e. Failure to check safe conditions during the operation or process
2. Supervisor did not plan the activity properly
   a. No planned safety, causing unsafe methods
   b. Using unskilled people that do not know safe operations
   c. No enforcement of discipline or safety rules
   d. Insufficient manpower, leading to people trying to do too much
3. Wrong design, construction, and bad housekeeping
   a. Poor layout of facility
   b. Faulty equipment
   c. Poor housekeeping, causing bad ventilation, insufficient lighting, buildup of rubbish
4. Failure to follow instructions and rules
   a. Safe work procedure and practices not followed
   b. Working without permission or authority
   c. Safety devices not inspected
5. Malfunctioning or absence of protective devices
   a. No protective device supplied
   b. Inadequate protection given by device
6. Not using the safety devices provided
   a. Inadequate enforcement to use protective devices, especially hearing protection devices
7. Specific physical condition
   a. Person is ill
   b. Lack of skill or physical ability

8. Lack of knowledge and mental ability
   a. Failure to concentrate
   b. Lack of good judgment
   c. Temper, anger, and impulsive actions
   d. Unnecessary haste and hazardous shortcuts
9. Use of devices with unknown defects
   a. Failure of material that cannot be foreseen
10. Accidents that cannot be foreseen
    a. Vandalism and industrial sabotage
    b. Wild animals, birds, and insects
    c. Acts of nature such as lightning

The number of people who died in America where death was attributed to occupational hazards has declined greatly over the past 20 years. In 1980 the number was 7,343 deaths—almost 21 people per day. In 1995 this number decreased to 5,314 people per year, or 15 per day. A preliminary total of 4,547 fatal work injuries were recorded in the United States in 2010. The reduction in the number of deaths is because of better management of safety.

15.3 Safety

Safety in general is a way of life. People that are prone to accidents might just not have the same accident prevention instincts that other people have. Public safety is certainly a concern and that is why the company ensures that the products that it sells will be good and will not cause illness and death.

The safety as discussed in this chapter deals with occupational safety. The preamble on the OSHA web page reads as follows: "A safe workplace is central to our ability to enjoy health, security, and the opportunity to achieve the American dream. Accordingly, assuring worker safety in a complex and sometimes dangerous modern economy is a vital function of our government." The reduction in the number of deaths is partly because of the commitment of OSHA to "assure so far as possible [for] every working man and woman in the nation safe and healthful working conditions."

Many of the sections that follow have been copied from the OSHA* web pages and are set apart from text as extracts. I am doing this because I might change the meaning of the thoughts even slightly, thereby changing the original intent.

---

* http://www.osha-slc.gov/html/Reinventing/index.html

15.3.1 Hazards in the Food Industry

There are fewer hazards in the food processing industry than in the chemical industry. It is, however, also true that someone working in a badly designed and poorly kept grain-handling facility is working inside a bomb waiting for the dust explosion to happen.

Fire in warehouses and dry material storage areas is a constant threat. Oil-soaked rags in a vehicle service area can spontaneously combust. People smoking in nondesignated smoke areas can also cause fires since they have to get rid of the smoking cigarette quickly when someone approaches.

In facilities where ammonia refrigeration is used, the possibility of an ammonia explosion exists. Just a small leak can cause serious breathing difficulties and panic. High-pressure steam is lethal when all that energy is released when a connection fails.

15.3.2 Handling Materials

- **Use of mechanical equipment.** Where mechanical handling equipment is used, sufficient safe clearances shall be allowed for aisles, at loading docks, through doorways, and wherever turns or passage must be made. Aisles and passageways shall be kept clear and in good repair, with no obstruction across or in aisles that could create a hazard. Permanent aisles and passageways shall be appropriately marked.
- **Secure storage.** Storage of material shall not create a hazard. Bags, containers, bundles, etc. stored in tiers shall be stacked, blocked, interlocked, and limited in height so that they are stable and secure against sliding or collapse.
- **Housekeeping.** Storage areas shall be kept free from accumulation of materials that constitute hazards from tripping, fire, explosion, or pest harborage. Vegetation control will be exercised when necessary.
- **Clearance limits.** Clearance signs to warn of clearance limits shall be provided.
- **Rolling railroad cars.** Derail and/or bumper blocks shall be provided on spur railroad tracks where a rolling car could contact other cars being worked on, or enter a building, work, or traffic area.
- **Guarding.** Covers and/or guardrails shall be provided to protect personnel from the hazards of open pits, tanks, vats, ditches, etc.

15.4 Accident Prevention

Safety has to do with making sure that people realize that a specific hazard exists and that they will be more careful to avoid changing the hazardous circumstance into an accident. One of the ways to do this is by color coding the workplace.

15.4.1 Color Coding

Color coding is effective to enhance safety in the workplace and should be used in combination with clearly printed labels. The color codes are not standardized for all industries and the meaning of the code used by the company should be displayed in all the buildings on the site (Table 15.1).

The general system that public works departments use to color code pipes and cables that are buried is given in Table 15.2.

Table 15.1 Standard Color-Code System to Identify Specific Hazards

Color	Related Hazard
Red	Fire-fighting equipment; containers for flammable and combustible[a] materials, flashpoint <27°C; emergency stop buttons or switches
Orange	Dangerous parts of machinery that can cut, crush, or cause some damage; all machine guards and inside of fuse boxes
Yellow	Physical dangers such as slippery areas, low beams that you can bump against, curbs, aisles, change in floor levels
Green	Safe areas, first-aid stations, and materials including gas masks
Blue	Caution machine under repair: "Out of order," Do not operate," and "Do not move" painted in white lettering
Purple	Radiation hazards
Black	Traffic control and general marking
White	Traffic control and general marking

[a] Combustible liquid means any liquid having a flash point at or above 60°C and below 93.4°C.

Table 15.2 Color Codes for Buried Pipes and Cables

Color	Used for
Red	Electric power lines, lighting cables, conduit
Yellow	Gas, oil, steam, petroleum
Orange	Communications cables, alarm cables, signal lines
Blue	Potable water, irrigation water, slurry lines
Green	Sewers, drain lines
Pink	Temporary survey markings
White	Proposed excavation

Table 15.3   Color Scheme for Pipes

Color	Contents
Red with white lettering	Fire protection, water, CO_2, foam, halon
Red with black lettering	Fuel gas pipes, gasoline, acetylene
Green with black lettering	All water pipes and liquid gas pipes (e.g., oxygen)
Yellow with red lettering	Natural gas, steam at 100°C or higher, other hazardous and flammable materials and high-pressure fluids
Yellow with black lettering	Ammonia
Blue with white lettering	Compressed air and low-pressure inert gases
Black bands	Vent lines
Gray bands	Service water and sprinkler pipes
Orange	Electrical conduit

Pipes in the processing area should also be color coded with a standard coding scheme that is used in the whole system (Table 15.3). For a company with multiple factories, it is a good idea to standardize color coding for all its factories. The safety officer is responsible for ensuring that everyone knows the meaning of the color codes.

Trying to identify every different line with a different color is rather stupid. Remember that many people are color blind, so the pipes have to be labeled. The other problem is that you run out of strong colors and have to rely on shades of color or you have to use stripes, which are much more difficult to paint than a solid color.

15.4.2 Operating Instructions

Good, clear instructions about the operation of a machine are essential for the safety of the operator and the safety of the process. The instructions should include a section regarding what to do in case of a problem. Operator alertness is the first defense to avoid out-of-control situations and accidents.

Instructions and training should emphasize the importance of preventative maintenance of safety devices and alarm systems. Competent inspections of these systems should be part of the housekeeping schedule. Safety can never be compromised just to save costs.

15.4.3 Pressure Equipment

Failure of pressure equipment can be prevented with the installation of an adequate number of relief valves. The valves must be checked on a regular basis and be protected against corrosion and fouling. Excessive temperatures are not a significant safety hazard but can harm the product, so automatic temperature limit controls should be checked. If the increase in temperature is caused by a simultaneous increase in pressure, it could mean that a pressure-limiting device is not working properly. In this case, a hazardous situation could be imminent.

All tanks and systems should be thoroughly purged before they are opened or entered for cleaning, maintenance, or inspection. It is vital to ensure that enough oxygen is present inside the tank. The fermentation industry has a particularly poor record with death caused by asphyxiation when fermentation vessels are cleaned.

15.4.4 Grain Handling Equipment

A dust explosion in an elevator is one of the most devastating food industry incidents. In dusty areas such as grain handling and storage facilities, the dust mixed with air forms an explosive mixture that requires a very small spark to set off the explosion. The real problem is that a small primary explosion can cause a shockwave that will make more dust airborne. The second explosion with its associated fireball is responsible for the terrible injuries and destruction of the facility.

The best way to avoid a dust explosion is to prevent the buildup of dust. Thorough housekeeping is essential. The amount of dust inside the handling equipment and silos is an unavoidable problem and ensuring the absence of sparks can prevent the explosion.

15.4.5 Flammable Material

Explosions of flammable materials can occur when discharge of static electricity causes a spark. Static electricity can be generated by surface friction between a conductor and a nonconductor, by friction between similar surfaces when there is relative motion of surfaces, and by turbulence of dust or liquid droplet-laden gases. Moving belts or vehicles, flowing gases or liquids, moving fluidized solids, and liquid atomization can generate static electricity.

Protection against this hazard requires that the potential difference should be reduced without a spark. This will work just like a lightning rod. Metal parts of vessels and all connecting pipes should be electrically grounded. Any tank truck or rail cars handling flammable materials should also be grounded.

15.5 General Plant Safety

Local and state legislation will set the primary safety regulations pertaining to building and plumbing codes. Insurance companies can also have their own codes from which they will determine their risk and therefore the premiums for the company. There are also the generally accepted safety practices that are based on experience.

In multistory buildings, an alternate means of descent should be provided from elevated levels. This could be ladders, outside staircases, or even a slide pole. All emergency routes should lead to an area that is safe. Any escape routes that end in hazardous areas or in a dead end should be replanned. It is normal practice to have an escape so that any person will not need to travel more than 12 m to get to an exit.

15.5.1 Inspection of Equipment

If equipment requires frequent inspection, it should be readily accessible. If the inspection has to be carried out during production, the access to the inspection point should be designed to ensure no contamination of the food product. When the inspection point is more than 2.4 m above the ground level, a special platform should be provided. The platform should have guardrails around it. Guardrails should also be installed around inspection pits and below-grade pumps.

15.5.2 Power Drives

All power-driven equipment such as belt drives, chain drives, and conveyors require suitable guards around them. All moving parts of machines are normally covered in such a way that even the most thoughtless action from an employee cannot cause a time-lost accident. It is good to remember that the frequency of the alternating current causes lights to act like a stroboscope. If a pulley rotates such that a specific point is visible in a multiple of the alternating current, the pulley would seem to be motionless. If a person tried to feel if the pulley is moving, the action could cause the loss of a finger. The guards need to be designed so that it is impossible to stick a finger through the guard and touch a moving part.

When equipment is inspected or repaired, a lock needs to be provided to ensure that the machine cannot be started. This is an essential safety issue and noncompliance can result in a fine. A horrible accident involved someone that was busy replenishing the graphite layer on a metal forming press. While he was busy, the press came down with fatal consequences. Accidents involving moving machinery are not forgiving of thoughtlessness.

Any processing vessel that operates at a temperature above 80°C should be insulated to prevent personal injury. The same is true for pipes carrying steam or hot fluids. They also need to be insulated to reduce heat loss to the environment that will cause additional ventilation cost to keep the area within comfortable temperature ranges. Where insulation is impractical, guards should be installed.

15.6 Fire Protection

Workplace fires and explosions kill 200 and injure more than 5,000 workers each year. In 1995, more than 75,000 workplace fires cost businesses more than $2.3 billion.

Fire protection is an important part of the overall operation of any food processing plant. Combustion is a process of self-contained exothermic oxidation (something burns and gives off heat). When any source of fuel is combined with air at the right temperature, self-ignition occurs. A spark or flame can cause a localized area of high temperature. In the case of dust explosions, the mixture is so unstable that combustion will start at ambient temperatures.

Three major types of fires occur in the food industry:

1. Combustible materials such as paper, wax, and raw materials like grain
2. Flammable* liquids such as oil, solvents, and petrochemicals
3. Electrical fires

To prevent fires and limit fire-related accidents, OSHA† recommends the following:

1. Eliminate fire hazards: keeping workspaces free of waste paper and other combustibles, replacing damaged electrical cords, and avoiding overloaded circuits
2. Prepare for emergencies: making sure all smoke detectors work, knowing who to call in an emergency, and participating in fire drills
3. Report fires and emergencies promptly: sounding the fire alarm and calling the fire department
4. Evacuate safely: leaving the area quickly in an emergency, using stairs instead of the elevator, and helping your co-workers

Identifying ignition hazards and controlling actions in these hazardous areas is a way to prevent fire. This will include control of smoking with notices posted, such as "No open flame" or "No smoking." Special sealed lights should be used in areas where flammable liquids and gases are stored. Make sure that the hot exhaust gases of internal combustion powered equipment are well away from combustible materials.

During an OSHA inspection the following are inspected:

---

* Please make sure to explain to foreign workers that inflammable means flammable and is different from other words starting with "in," where it means "not," such as insignificant and insincere.
† http://www.osha-slc.gov/OshDoc/Fact_Data/FSN093-41.html

- Building fire exits
  - Each workplace building must have at least two means of escape remote from each other to be used in a fire emergency.
  - Fire doors must not be blocked or locked to prevent emergency use when employees are within the buildings. Delayed opening of fire doors is permitted when an approved alarm system is integrated into the fire door design.
  - Exit routes from buildings must be clear and free of obstructions and properly marked with signs designating exits from the building.
- Portable fire extinguishers
  - Each workplace building must have a full complement of the proper type of fire extinguisher for the fire hazards present, excepting when employers wish to have employees evacuate instead of fighting small fires.
  - Employees expected or anticipated to use fire extinguishers must be instructed on the hazards of fighting fire, how to properly operate the fire extinguishers available, and what procedures to follow in alerting others to the fire emergency.
  - Only approved fire extinguishers are permitted to be used in workplaces, and they must be kept in good operating condition. Proper maintenance and inspection of this equipment is required of each employer.
  - Where the employer wishes to evacuate employees instead of having them fight small fires there must be written emergency plans and employee training for proper evacuation.
- Emergency evacuation planning
  - Emergency action plans are required to describe the routes to use and procedures to be followed by employees. Also, procedures for accounting for all evacuated employees must be part of the plan. The written plan must be available for employee review.
  - Where needed, special procedures for helping physically impaired employees must be addressed in the plan; also, the plan must include procedures for those employees who must remain behind temporarily to shut down critical plant equipment before they evacuate.
  - The preferred means of alerting employees to a fire emergency must be part of the plan and an employee alarm system must be available throughout the workplace complex and must be used for emergency alerting for evacuation. The alarm system may be voice communication or sound signals such as bells, whistles, or horns. Employees must know the evacuation signal.
  - Training of all employees in what is to be done in an emergency is required. Employers must review the plan with newly assigned employees, so they know correct actions in an emergency, and with all employees when the plan is changed.

■ Fire prevention plan
  - Employers need to implement a written fire prevention plan to complement the fire evacuation plan to minimize the frequency of evacuation. Stopping unwanted fires from occurring is the most efficient way to handle them. The written plan shall be available for employee review.
  - Housekeeping procedures for storage and cleanup of flammable materials and flammable waste must be included in the plan. Recycling of flammable waste such as paper is encouraged; however, handling and packaging procedures must be included in the plan.
  - Procedures for controlling workplace ignition sources such as smoking, welding, and burning must be addressed in the plan. Heat producing equipment such as burners, heat exchangers, boilers, ovens, stoves, fryers, etc. must be properly maintained and kept clean of accumulations of flammable residues; flammables are not to be stored close to these pieces of equipment.
  - All employees are to be apprised of the potential fire hazards of their job and the procedures called for in the employer's fire prevention plan. The plan shall be reviewed with all new employees when they begin their job and with all employees when the plan is changed.
■ Fire suppression system
  - Properly designed and installed fixed fire suppression systems enhance fire safety in the workplace. Automatic sprinkler systems throughout the workplace are among the most reliable fire fighting means. The fire sprinkler system detects the fire, sounds an alarm, and puts the water where the fire and heat are located.
  - Automatic fire suppression systems require proper maintenance to keep them in serviceable condition. When it is necessary to take a fire suppression system out of service while business continues, the employer must temporarily substitute a fire watch of trained employees standing by to respond quickly to any fire emergency in the normally protected area. The fire watch must interface with the employer's fire prevention plan and emergency action plan.
  - Signs must be posted about areas protected by total flooding fire suppression systems, which use agents that are a serious health hazard, such as carbon dioxide, Halon 1211, etc. Such automatic systems must be equipped with area predischarge alarm systems to warn employees of the impending discharge of the system and allow time to evacuate the area. There must be an emergency action plan to provide for the safe evacuation of employees from within the protected area. Such plans are to be part of the overall evacuation plan for the workplace facility.

15.6.1 Fire Fighting

Fires can be extinguished if the source of combustible materials is removed, such as shutting a valve on an oil line. Fires can also be doused by removing the source of oxygen or by lowering the temperature below the firing temperature.

When water is sprayed on a fire, it cools down the fire and forms steam, which will expand rapidly to replace the airflow that supplies the oxygen. Spraying with carbon dioxide foam removes the source of oxygen and cools the fire down. Using sand will limit the flow of oxygen and starve the fire.

Two types of foam are in general use. The first is foam consisting of CO_2 bubbles in water. The second is mechanically produced foam where air is beaten into water that contains a foam-producing substance. The foams are very effective since they adhere to solids.

Water is an efficient coolant and will serve to blanket heavy liquids. Water has minimal cooling effect on burning light liquids. Fogging with high-pressure nozzles enhances vaporization and the cooling effect.

When fighting electrical fires, precautions must be taken not to electrocute the firefighter. Water and fog should be avoided on electrical fires since the potential of electrical shock is great. CO_2 and dry powders are effective against electrical fires and flammable liquids.

Small fires are normally extinguished with a handheld extinguisher. It is important that the right kind of extinguisher be used and that the operator knows how to use it. Small fires can very quickly get out of control when an incompetent person uses all the extinguishers incorrectly. It is best to train the fire-fighting team and, in the case of a fire, everyone else should get out of harm's way.

Water sprinkler systems that are normally automatically controlled are the most widely used systems for control of building fires. Sprinkler systems are also very dependable since their jets are larger and do not clog easily. When a fogging spray is used, the volume of water is reduced while the cooling effect is maximized. This leads to less water damage of the equipment.

15.6.1.1 Portable Extinguishers

All portable fire extinguishers are coded from type A to type D to ensure that they are used for the right purpose. For a fire involving ordinary combustibles, type A should be used. Type B extinguishers are used for fires involving liquids, grease, and gases. For electrical fires, type C should be used; type D units are used for metals such as sodium, magnesium, potassium, etc.

Letter symbol markings are prominently displayed on the extinguishers. When an extinguisher is of a combined type, the letter symbols should be placed in a horizontal sequence. Type A and B units also have a numerical rating that indicates the size of the fire they can control. A 6-A unit will control a fire three times as large as a 2-A unit.

15.7 Noise Control

Noise* as a cause of hearing loss has led to many disability claims. This is a very complex issue in a court of law. The fact that the claimant has been hunting without hearing protectors and that most of the hearing loss can be attributed to noise from non-work-related sources does nothing to diminish the settlement amounts. OSHA has specific noise standards and it is best to ensure that the safe levels are not exceeded.

In one industry where the testing laboratory was particularly noisy, management reserved the positions for people who were legally deaf. With sign language, they could communicate and work without being encumbered with the bulky noise suppressors that would otherwise have been required. The company had effectively reduced its liability for hearing loss and enhanced its status as being disabled friendly.

Food processing plants are noisy enough to make normal conversation difficult, but not so noisy that hearing loss is a serious risk. It is, however, very important that the noise level be checked under full different production capacities to ensure that safe noise levels are never exceeded.

Noise and sound levels are measured in decibels (dB). The intensity of sound is measured on a scale starting at one for the least perceptible sound and increasing with multiplying intensity for each decibel rise. Noise becomes painful at about the 120 dB level.

The decibel scale is logarithmic and should be interpreted as such. The threshold of hearing is 10^{-12} W/m². At an intensity of 1 W/m², the sound will become painful and this level is called the threshold of pain. For our perception, a sound with an intensity of 10^{-5} W/m² will be about twice as loud as one with an intensity of 10^{-6} W/m².

By using the threshold of hearing (I_0) as the base and the sound intensity (I) as numerator, we can calculate the logarithmic ratio as follows:

$$\log \frac{I}{I_0} = \log \frac{10^{-5} \text{ W/m}^2}{10^{-12} \text{ W/m}^2} = \log 10^7 = 7 \text{ B} = 70 \text{ dB}$$

Example 15.1

What is the difference in intensity of a sound when it starts at 40 dB and goes to 70 dB?

Since we are working with logarithms that are exponents of 10, we have to think of manipulations in terms of exponents. Subtracting exponents is the same as division of the numbers:

---

* Noise is a loud, disagreeable, nonmusical sound. Hearing loss occurs as readily by listening to music as by listening to noise. There is no distinction; both are loud sounds.

Table 15.4   Sound Source and Its Decibel Intensity

Sound Source	Decibel Level
Soft whisper	20
Normal conversation	60
City traffic	70
Average factory	80
Machine shop	100
Rock band with amplifiers	110
Pneumatic drill	120
Jet plane at takeoff	140

$$B = 10\log\frac{I_2}{I_1} = 70 - 40 = 30$$

$$\log\frac{I_2}{I_1} = 3$$

$$\frac{I_2}{I_1} = 10^3$$

$$I_2 = 1000 I_1$$

There is a 1000-fold increase in the sound intensity when we go from 40 to 70 dB (Table 15.4).

The EPA claims that noise levels at work should be reduced to less than 75 dB to prevent all workplace-related noise damage. This is too low for practical purposes and the usable standard of 85 dB is used. The length of time that a person is exposed to high noise levels is also important. An orchestra could play at an intensity of 100 dB for 2 h without permanent damage to the ear (Table 15.5).

According to OSHA, only 2 percent of workers could expect to lose hearing ability after 30 years' exposure to 90 dB of noise each day. The debate about the cause of hearing loss and liability litigation will go on forever.

Example 15.2

How much noise is created from two sources where one has an intensity of 70 dB and the other has an intensity of 80 dB?

Since we are working with logarithms, do not fall into the trap of simplistic calculations. Simple summation means that you are multiplying intensities.

Table 15.5   Permissible Noise Exposures

Duration of Noise per Day (h)	Sound Level (dB)
8	90
6	92
4	95
3	97
2	100
1.5	102
1	105
0.5	110
0.25 or less	115

$$\beta_1 = 70 \text{ dB} = 10\log\frac{I_1}{I_0} = 10\log\frac{I_1}{10^{-12} \text{ W/m}^2} = 70 \text{ dB} = 10\log 10^{-7}$$

$$I_1 = 10^{-5} \text{ W/m}^2$$

$$\beta_2 = 80 \text{ dB} = 10\log\frac{I_2}{I_0} = 10\log\frac{I_2}{10^{-12} \text{ W/m}^2} = 80 \text{ dB} = 10\log 10^{-8}$$

$$I_2 = 10^{-4} \text{ W/m}^2$$

The total noise intensity is the sum of the two intensities:

$$I_{total} = I_1 + I_2 = 1\times10^{-5} + 1\times10^{-4} = 1\times10^{-5} + 10\times10^{-5} = 11\times10^{-5}$$

$$\beta = \log\frac{I_{total}}{I_0} = 10\log\left(\frac{11\times10^{-5} \text{ W/m}^2}{10^{-12} \text{ W/m}^2}\right) = 10\log(11\times10^7)$$

$$\beta = 10\log(1.041 + 7.0) = 80.41 \text{ dB}$$

This is much different from the intuitive response of a sound much larger than 80 dB.

15.8 Occupational Safety and Health Administration

The Occupational Safety and Health Administration caused a very significant change in the design and operation of all food plants. The act involves management and every individual worker and it applies to all processing companies that have one or more employees. It was originally intended to remove health hazards from industry. One of the problems that arises is when suicide is attributed to a circumstance at work. It then becomes an accident that should have been prevented!

The OSHA standards are a list of rules intended to safeguard the worker by listing equipment and practices associated with accident prevention. Whenever an employee requests an inspection of a workplace that is considered unsafe, OSHA will investigate without delay. For the most part, the act has a voluntary principle until something goes wrong. At that point, an inspection could point out many serious failures to comply with concomitant penalties. The fines are large enough to be a strong motivational factor to comply with the regulations. Employers must maintain records of each job-related accident or illness, and these must be available for inspection.

> The number of focused inspections rose from 377 in the first half of Fiscal Year 1995 to 996 in the second half of the year. We are convinced that this approach is an effective way (1) to promote the development and implementation of effective worksite safety and health programs and (2) to target the most hazardous operations on a construction worksite. We are developing a focused inspection policy for general industry. At the same time, OSHA continues to enforce the law vigorously against those employers who ignore their responsibilities under law. In Fiscal Year 1995, for example, OSHA issued initial penalties of $100,000 or more against 122 employers compared with 68 such penalties in Fiscal Year 1994.

In the situation of a unionized workforce, the union cannot argue the case of an employee that was dismissed by an employer if the employee is proven to be in violation of a standard—especially if it can be proved that he or she was given a fair and equitable opportunity to mend his or her ways.

An individual, normally a safety director, will be responsible to ensure that the company complies with the act and that a safety committee is organized to get the job done.

Bibliography

Brain Bank. Environmental issues—Hazardous waste. Available at http://www.cftech.com/BrainBank/MANUFACTURING/EnvironIss.html

Brennan, J. G., Butters, J. R., and Cowell, N. D. 1990. *Food engineering operations,* 3rd ed. London: Elsevier Applied Science.

Burton, R. M., Chandler, J. S., and Holtzer, H. P. 1986. *Quantitative approaches to business decision making.* New York: Harper & Row.

Cleaver-Brooks. Boiler Selection Guide. Available at http://www.cleaver-brooks.com/Boilersa2.html

Cumberland, R., for C. P. I. Equipment Ltd., Waterworks Warehouse. Ion exchange. Available at http://www.netos.com/cpi/ion_exch.htm

Danfoss. Refrigeration and air conditioning controls. Available at http://www.danfoss.com/Business/a121ref.htm

Debnath, L. (2006). A brief historical introduction to fractals and fractal geometry. *International Journal of Mathematical Education in Science & Technology* 37 (1): 29–50.

Diversified Panel Systems, Inc. Manufacturers of quality machine laminated insulated building materials. Available at http://www.doublebarrel.com/DPS/DPSproducts.htm#top

Farrall, A. W. 1979. *Food engineering systems,* vol. 2. Westport, CT: AVI Publishing Co.

Faulkner, V. N. (2009). The components of number sense. *Teaching Exceptional Children* 41 (5): 24–30.

Fellows, P. 1988. *Food processing technology, principles and practice.* Cambridge: Woodhead Publishing Ltd.

Fiorentin, S. (2009). A re-interpretation of the concept of mass and of the relativistic mass–energy relation. *Foundations of Physics* 39 (12): 1394. doi:10.1007/s10701-009-9359-9

Fogarty, D. W., Hoffmann, T. R., and Stonebraker, P. W. 1989. *Production and operations management.* Cincinnati, OH: South Western Publishing Co.

Harper, J. C. 1976. *Elements of food engineering.* Westport, CT: AVI Publishing Co.

Hartel, R. W., Howell, T. A., and Hyslop, D. B. 1997. *Math concepts for food engineering.* Lancaster, PA: Technomic Publishing Company, Inc.

Hayes, G. D. 1987. *Food engineering data handbook.* New York: Longman Scientific and Technical.

Henderson, S. M., Perry, R. L. and Young, J. H. 1997. *Principles of process engineering,* 4th ed. St. Joseph, MI: ASAE.

ICI. KLEA™ physical property summary. Available at http://www.dircom.co.uk/klea/rppk134.html

Kamp, P. (2011). The one-second war. *Communications of the ACM* 54 (5): 44–48. doi:10.1145/1941487.1941505

Keighley, H. J. P. 1986. *Work out physics.* Hampshire: Macmillan Education Ltd.

Lee, S. (2011). Scalar-tensor theory in ten dimensions with anisotropic matter. *Modern Physics Letters A* 26 (1): 19–30.

Lewis, M. J. 1987. *Physical properties of foods and food processing systems.* West Sussex: Ellis Horwood Ltd.

Memcor Application Bulletin A1002. CMF application parameters—Industrial waste water. Available at http://www.memtec.com/memcor/techinfo/a1002.html

Ministry of Environment and Energy, Denmark. Environmental project no. 301, 1995. Going towards natural refrigerants. Available at http://www.mem.dk/mst/pubs/no301/Summary.htm

Perry, R. H., and Green, D. W. 1997. *Perry's chemical engineers' handbook,* 7th ed. New York: McGraw–Hill.

ResinTech, Inc. Typical properties of cation exchange resins. Available at http://www.resin-tech.com/prop.htm#ATION

Rogers, J. P. (2008). Cardinal number and its representation: Skills, concepts and contexts. *Early Child Development & Care* 178 (2): 211–225. doi:10.1080/03004430600722705

Rohmhaas. Ion exchange resins. Available at http://www.rohmhaas.com/businesses.dir/IonEx.html

South African Pump Manufacturers Association. 1993. K. Myles and Associates, Northcliff.

Spirax Sarco. Steam traps. Available at http://www.spirax-sarco.com.au/spirax.htm

Toledo, R. T. 1994. *Fundamentals of food process engineering.* New York: Chapman & Hall.

US Department of Energy. Federal technology alert refrigerant subcooling. Available at http://www.pnl.gov/fta/1_refrig.htm

Wong Brothers Refrigeration SDN. BHD. Panels Manufacturing Division. Available at http://www.geocities.com/Eureka/Park/2368/

Appendix 1

A1.1 Numerical Prefixes of SI Units

Factor	Scientific	In Words	SI Prefix	SI Symbol
1,000,000,000	10^9	Billion/milliard/trillion	Giga	G
1,000,000	10^6	Million	Mega	M
1,000	10^3	Thousand	Kilo	k
100	10^2	Hundred	Hecto	h
10	10	Ten	Deca	da
0.1	10^{-1}	Tenth	Deci	d
0.01	10^{-2}	Hundredth	Centi	c
0.001	10^{-3}	Thousandth	Milli	m
0.000001	10^{-6}	Millionth	Micro	μ
0.000000001	10^{-9}	Billionth/milliardth	Nano	n

A1.2 Conversion Tables

The following conversion tables will provide conversion between SI, metric, US, and imperial systems. All the tables use a multiplying factor.

A1.2.1 Length

From To	Millimeter	Centimeter	Meter	Kilometer	Inch	Foot	Yard	Mile
Millimeter	1	0.1	0.001	—	0.03937	—	—	—
Centimeter	10	1	0.01	—	0.393701	0.032808	—	—
Meter	1000	100	1	0.001	39.3701	3.28084	1.09361	—
Kilometer	—	—	1000	1	—	3280.84	1093.61	0.621371
Inch	25.4	2.54	—	—	1	0.083333	0.02778	—
Foot	304.8	30.48	0.3048	—	12	1	0.33333	—
Yard	914.4	91.44	0.9144	0.000914	36	3	1	0.000568
Mile	—	—	1609.344	1.609344	—	5280	1760	1

A1.2.2 Area

From To	cm^2	m^2	km^2	$in.^2$	ft^2	yd^2	acre	$mile^2$
cm^2	1	0.0001	—	0.155	0.001076	0.0001196	—	—
m^2	10000	1	0.000001	1550	10.7639	1.19599	0.0002471	—
km^2	—	1000000	1	—	—	—	247.105	0.386102
$in.^2$	6.4516	0.000645	—	1	0.006944	0.000772	—	—
ft^2	929.03	0.092903	—	144	1	0.111111	0.000023	—
yd^2	8361.27	0.836127	—	1296	9	1	0.0002066	—
acre	—	4046.86	0.004047	43560	4840	1	0.001562	
$mile^2$	—	—	2.589987	—	—	—	640	1

A1.2.3 Mass

From To	kg	ton	lb	UK cwt.	UK ton	US cwt.	US ton
kg	1	0.001	2.20462	0.019684	0.000984	0.022046	0.001102
ton	1000	1	2204.62	19.6841	0.984207	22.0462	1.10231
lb	0.453592	0.000454	1	0.008929	0.000446	0.01	0.0005
UK cwt.	50.8023	0.050802	112	1	0.05	1.12	0.056
UK ton	1016.05	1.01605	2240	20	1	22.4	1.12

From To	kg	ton	lb	UK cwt.	UK ton	US cwt.	US ton
US cwt.	45.3592	0.045359	100	0.892857	0.044643	1	0.03
US ton	907.185	0.907185	2000	17.8517	0.892857	20	1

A1.2.4 Volume and Capacity

From To	cm³	m³	liter (dm³)	ft³	yd³	UK pint	UK gal	US pint	US gal
cm³	1	—	0.001	0.00004	—	0.00176	0.00022	0.00211	0.00026
m³	—	1	1000	35.3147	1.30795	1759.75	219.969	2113.38	264.172
liter	1000	0.001	1	0.03532	0.00131	1.75975	0.21997	2.11338	0.26417
ft³	28316.8	0.02832	28.3168	1	0.03704	49.8307	6.22883	59.8442	7.48052
yd³	764555	0.76456	764.555	27	1	1345.43	168.178	1615.79	201.974
UK pint	568.261	0.00057	0.56826	0.02007	0.00074	1	0.125	1.20095	0.15012
UK gal	4546.09	0.00455	4.54609	0.16054	0.00595	8	1	9.6076	1.20095
US pint	473.176	0.00047	0.47318	0.01671	0.00062	0.83267	0.10408	1	0.125
US gal	3785.41	0.00379	3.78541	0.13368	0.00495	6.66139	0.83267	8	1

A1.2.5 Pressure

From To	atm	mm Hg	m bar	bar	pascal	in H₂O	in Hg	psi
atm	1	760	1013.25	1.0132	101325	406.781	29.9213	14.6959
mm Hg	0.0013158	1	1.33322	0.001333	133.322	0.53524	0.03937	0.019337
m bar	0.0009869	0.75006	1	0.001	100	0.401463	0.02953	0.014504
bar	0.9869	750.062	1000	1	100000	401.463	29.53	14.504
pascal	0.0000099	0.007501	0.01	0.00001	1	0.004015	0.0002953	0.000145
in H₂0	0.0024583	1.86832	2.49089	0.002491	249.089	1	0.0736	0.03613
in Hg	0.033421	25.4	33.8639	0.0338639	3386.39	13.5951	1	0.491154
psi	0.068046	51.7149	68.9476	0.068948	6894.76	27.6799	2.03602	1

A1.2.6 Volume Rate of Flow

From To	L/h	m³/s	m³/h	ft³/h	UK gal/m	UK gal/h	US gal/m	US gal/h
L/s	3600	0.001	3.6	127.133	13.198	791.888	15.8503	951.019
L/h	1	—	0.001	0.03535	0.003666	0.219979	0.00440	0.264172
m³/s	3600000	1	3600	127133	13198.1	791889	15850.3	951019
m³/h	1000	0.000278	1	35.3147	3.66615	219.969	4.40286	264.1718
ft³/h	28.3168	—	0.028317	1	0.103814	6.22883	0.12468	7.480517
UK gal/m	272.766	0.000076	0.272766	9.63262	1	60	1.20095	72.057
UK gal/h	4.54609	—	0.004546	0.160544	0.016667	1	0.02002	1.20095
US gal/m	227.125	0.000063	0.227125	8.020832	0.832674	49.96045	1	60
US gal/h	3.785411	—	0.003785	0.133681	0.013368	0.83267	0.016667	1

A1.2.7 Power

From To	Btu/h	w	Kcal/h	KW
Btu/h	1	0.293071	0.251996	0.000293
W	3.41214	1	0.859845	0.001
Kcal/h	3.96832	1.163	1	0.001163
KW	3412.14	1000	859.845	1

A1.2.8 Energy

From To	Btu	Therm	J	kJ	Cal
Btu	1	0.00001	1055.06	1.055	251.996
Therm	100000	1	—	105500	25199600
J	0.00094	—	1	0.001	2388
kJ	0.9478	0.000009478	1000	1	238.85
Cal	0.0039683	0.0039683×10^{-5}	4.1868	—	1

A1.2.9 Specific Heat

From To	Btu/lb °F	J/kg °C
Btu/lb °F	1	4186.8
J/kg °C	0.00023	1

A1.2.10 Heat Flow Rate

From To	Btu/ft² h	W/m²	kcal/m² h
Btu/ft² h	1	3.154	2.712
W/m²	0.3169	1	0.859
kcal/m² h	0.368	1.163	1

A1.2.11 Thermal Conductance

From To	Btu/ft² h °F	W/m² °C	kcal/m² h/°C
Btu/ft² h °F	1	5.67826	4.88243
W/m² °C	0.176110	1	0.859845
kcal/m² h/°C	0.204816	1.163	1

A1.2.12 Heat per Unit Mass

From To	Btu/lb	kJ/kg
Btu/lb	1	2.326
kJ/kg	0.4299	1

A1.2.13 Linear Velocity

From To	ft/min	ft/s	m/s
ft/min	1	0.016666	0.00508
ft/s	60	1	0.3048
m/s	196.850	3.28084	1

Temperature conversion can be achieved by using the following formula

°F to °C: °C = (°F − 32) × 5/9
°C to °F: °F = (°C × 9/5) + 32

Appendix 2

A2.1 Some Help with Calculations

What follows are some basic rules of calculations. Most readers will find them familiar; however, this section is included for those that will benefit from a general refresher. Included is a small section outlining the rules applying to exponents and fractions. Throughout many years of helping many students with mathematical problems, I learned that a lack of understanding fractions and exponents is the base cause for a dislike of things numerical.

A2.2 The Plus/Minus Rule

A number or symbol may be moved from one side of an equation to the other side only if the sign in front of the number is changed. A "plus" item on one side of the equation will become a "minus" item on the other side of the equation and vice versa.

$$x = y + 5 \text{ or } x - y = 5 \text{ or } x - 5 = y$$

Using the "do on both sides" rule, one can write the equation as

$$x = y + 5$$

$$x - 5 = y + 5 - 5$$

$$x - 5 = y$$

A2.3 Fractions

A2.3.1 The Diagonal Rule

An item in a fraction may be moved diagonally across an equal sign

$$\frac{A}{B} = \frac{C}{D}$$

can be written as

$$AD = BC$$

and also as

$$\frac{A}{C} = \frac{B}{D} \quad \text{or} \quad \frac{D}{B} = \frac{C}{A}$$

Let us do it more slowly:

$$AD = BC$$

$$\frac{AD}{D} = \frac{BC}{D}$$

$$A = \frac{BC}{D}$$

$$\frac{A}{C} = \frac{BC}{DC} = \frac{B}{D}$$

A2.3.2 "Do on Both Sides" Rule

Whatever is done on one side of the equal sign must be repeated on the other side in order to maintain equality. If

$$A = B$$

then

$$2A = 2B \text{ (both sides were multiplied by 2)}$$

or

$$A^2 = B^2 \text{ (both sides were squared)}$$

If

$$X = \frac{1}{Y}$$

then

$$\frac{1}{X} = Y \text{ (both sides were inverted)}$$

If the symbols are confusing, replace them with simple numbers and see if the rule works. If

$$0.5 = \frac{1}{2}$$

then

$$\frac{1}{0.5} = 2$$

If

$$x = \frac{1}{y}$$

then we can multiply on both sides by 2 to get

$$2x = 2 \times \frac{1}{y} = \frac{2}{1} \times \frac{1}{y} = \frac{2}{y}$$

A mistake that students sometimes make is to multiply both the numerator and the denominator. If

$$\frac{1}{R} = \frac{1}{R_1} + \frac{1}{R_2}$$

then it cannot be said that

$$R = R_1 + R_2$$

If this is hard to see, try real numbers. Thus,

$$\frac{1}{2} = \frac{1}{3} + \frac{1}{6}$$

and $2 \neq 3 + 6$.

In this case, the whole item, $(1/3 + 1/6)$, must be inverted:

$$\frac{1}{R} = \frac{1}{R_1} + \frac{1}{R_2} = R = \frac{1}{\dfrac{1}{R_1} + \dfrac{1}{R_2}} = \frac{1}{\dfrac{R_2 + R_1}{R_1 R_2}} = \frac{R_1 R_2}{R_1 + R_2}$$

$1/2 = 1/3 + 1/6$, inverting both sides, gives

$$2 = \frac{1}{\dfrac{1}{3} + \dfrac{1}{6}} = \frac{1}{\dfrac{6+3}{3\times 6}} = \frac{3\times 6}{3+6} = \frac{18}{9} = 2$$

Did you notice what happened to the numerator and denominator when a fraction was inverted?

It looks far more complicated and difficult than what it is.

A2.3.3 Adding Fractions

Reviewing a very important rule will help to clarify the previous example. To add or subtract fractions, the denominators (the number below the line) must be the same. The quickest way to make them the same is to multiply the numerator (the number above the line) and the denominator of one fraction with the denominator of the other fraction:

$$\frac{x}{y} + \frac{a}{b} = \frac{xb}{yb} + \frac{ay}{yb} = \frac{xb+ay}{yb}$$

$$\frac{2}{3} + \frac{3}{4} = \frac{2\times 4 + 3\times 3}{3\times 4} = \frac{8+9}{12} = \frac{17}{12} = 1.417$$

$$\frac{2}{3} + \frac{3}{4} = 0.667 + 0.75 = 1.417$$

A2.3.4 Multiplying Fractions

When multiplying fractions, we multiply the denominators with each other and the numerators with each other:

$$\frac{x}{y}\times\frac{a}{b}\times c = \frac{x}{y}\times\frac{a}{b}\times\frac{c}{1} = \frac{xac}{yb}$$

$$\frac{2}{3}\times\frac{4}{5}\times 6 = \frac{2\times 4\times 6}{3\times 5} = \frac{48}{15} = 3\frac{3}{15} = 3\frac{1}{5} = 3.2$$

$$0.667\times 0.8\times 6 = 3.2$$

Always check your use of fractions and exponents by using simple numbers. The small amount of time invested can save a large amount of time wasted!

A2.3.5 Dividing Fractions

When dividing with a fraction, invert the numerator and use the multiplication rule:

$$\frac{a}{b}\div\frac{c}{d} = \frac{a}{b}\times\frac{d}{c} = \frac{ad}{bc}$$

$$\frac{3}{4}\div\frac{2}{5} = \frac{3}{4}\times\frac{5}{2} = \frac{15}{8} = 1.875$$

$$\frac{3}{4}\div\frac{2}{5} = \frac{0.75}{0.4} = 1.875$$

A2.4 Exponents

The mathematical principles used in working with fractions are similar to those for exponents. Remember when you work with exponents that anything with the exponent of zero is equal to one with the exception of zero itself. WE EXCLUDE ZERO FROM THIS DEFINITION SINCE IT IS MEANINGLESS AND IT WILL CAUSE HAVOC WITH ALL OUR EQUATIONS. 0^0 is an undefined quantity and we will just keep it as such. Zero with any exponent other than zero is zero. Be careful not to divide by zero because this causes all sorts of disarray in mathematics.[*]

What follows are some ground rules regarding exponents:

---

[*] If zero is a big problem in mathematics, and zero means nothing, then nothing is a big problem in mathematics.

1. A product of two numbers with an exponent can be written as

$$(xy)^a = x^a \times y^a$$

$$6^2 = (2 \times 3)^2 = 2^2 \times 3^2$$

This works both ways: If you have to multiply two numbers that have the same exponent, you can multiply the two numbers and the product will have the same exponent as the individual numbers.

2. Where we multiply numbers with different exponents but similar base numbers, we can simply add the exponents:

$$2x^3 = 2 \times x \times x \times x = 2 \times x^1 \times x^1 \times x^1 = 2x^{1+1+1} = 2x^3$$

2 is the coefficient, x is the base, and 3 is the exponent.

$$x^a \times x^b = x^{a+b}$$

$$2^2 \times 2^3 = 2^5$$

$$4 \times 8 = 32 = 2^5 = 2^2 \times 2^2 \times 2^2 \times 2^2 \times 2^2$$

If the exponents are not the same and the base numbers are not the same, multiplication is just not possible:

$$x^a \times y^b = x^a \times y^b$$

3. When an exponential number is raised to an exponent, we multiply the exponents:

$$(x^a)^b = x^{ab}$$

$$(2^3)^2 = 2^6$$

$$(2^3)^2 = (8)^2 = 64 = 2^6$$

The square root of any number is that number to the exponent of ½. In the same way, the fourth root of a number is that number to exponent ¼. The same rule applies:

$$\sqrt{x^a} = (x^a)^{1/2} = x^{a/2}$$

$$\sqrt{16} = \sqrt{4^2} = (4^2)^{1/2} = 4^{2/2} = 4^1 = 4$$

4. When we divide exponential numbers where the base numbers are the same, we get

$$\frac{x^a}{x^b} = x^a \times x^{-b} = x^{a+(-b)} = x^{a-b}$$

$$\frac{2^3}{2^2} = 2^{3-2} = 2$$

$$\frac{8}{4} = 2$$

Notice the change of sign on the exponent when we inverted the base number to do the multiplication.

When we divide exponential numbers with different base numbers, even if the exponents are the same x^a/b^a, they cannot be manipulated.

When you do try to do anything, strange answers may result. Use small numbers to check the rule:

$$\frac{2^2}{3^2} = \frac{4}{9} = 0.44$$

$$\frac{2^2}{3^2} = \left(\frac{2}{3}\right)^2 = 0.667^2 = 0.44$$

$$\frac{2^2}{3^2} \neq \frac{\sqrt{2^2}}{\sqrt{3^2}} = \frac{2}{3} = 0.667$$

With simple fractions, we can multiply and divide both the numerator and denominator with the same amount and the ratio will stay the same. This does not work when you manipulate exponents. Multiplying exponents means raising the exponential number by the power of the multiplier. Dividing exponents means taking roots of the numbers. Be very careful when you work with exponents.

A2.5 Quadratic Equations

Quadratic equations can be generalized as

$$ax^2 + bx + c = 0$$

where a, b, and c are numbers called coefficients. The equation can be solved by factorization, drawing a graph, or using a formula.

A2.5.1 Solution by Factorization

An easy test to determine if an equation will factorize is as follows:

Multiply the coefficients a and c.

Write down the factor pairs of the product.

If the sign of c is +, then adding one of the pairs together will give the coefficient b if the equation factorizes.

If the sign of c is −, then subtracting one of the pairs together will give the coefficient b if the equation factorizes.

The procedure can be illustrated using the equation

$$4x^2 + 5x - 6 = 0$$

The product ac = 24.

The factor pairs of 24 are (1,24) (2,12) (3,8) (4,6).

Since c is minus, one of these pairs must subtract to give the coefficient of x. The required pair is (3,8).

Writing the equation with this pair gives

$$4x^2 + 8x - 3x - 6 = 0$$

$$4x(x+2) - 3(x+2) = 0$$

$$(x+2)(4x-3) = 0$$

$$x + 2 = 0$$

$$x = -2$$

$$4x - 3 = 0$$

$$x = \frac{3}{4}$$

A2.5.2 Solution Using the Formula

If an equation does not factorize, the following formula can be used:

$$x = \frac{-b \pm \sqrt{b^2 - 4ac}}{2a}$$

Factorization by use of the formula can be illustrated with the equation $3x^2 - 4x - 1 = 0$. This equation does not factorize. The coefficients are $a = 3$, $b = -4$, and $c = -1$. Therefore,

$$x = \frac{-(-4) \pm \sqrt{\{(-4)^2 - 4(3)(-1)\}}}{2(3)}$$

$$x = \frac{4 \pm \sqrt{\{16 + 12\}}}{6} = \frac{4 \pm \sqrt{28}}{6}$$

$$x = \frac{4 \pm 5.291}{6}$$

$$x = \frac{9.291}{6} = 1.55$$

$$x = \frac{-1.291}{6} = -0.22$$

A2.5.3 Simultaneous Equations

An equation such as $y + x = 5$ has an infinite number of pairs that will satisfy it. This is also true for the equation $y - x = 3$. There are, however, only specific values for x and y that will satisfy both equations:

$$y + x = 5$$

$$y - x = 3$$

Adding the two equations gives

$$2y = 8$$

$$y = 4$$

$$x = 1$$

This is rather simplistic. The rule is that we need the same number of equations as we have unknowns. Working with three unknowns requires three equations and so forth:

$$x + 2y + 3z = 20$$

$$2x + y + z = 11$$

$$3x - 2y + 2z = 8$$

The easiest unknown to remove will be *y*. If we multiply the second equation by two, we can subtract that from the first equation to get

$$x + 2y + 3z = 20$$

$$4x + 2y + 2z = 22$$

$$-3x + z = -2$$

Adding equation 3 to equation 1 will also eliminate *y* and we get

$$x + 2y + 3z = 20$$

$$3x - 2y + 2z = 8$$

$$4x + 5z = 28$$

We are left with two equations and two unknowns. There is no easy way to get rid of either of the two unknowns, so we will multiply the first one with four and the second with three to get the *x* values the same:

$$-12x + 4z = -8$$

$$12x + 15z = 84$$

Adding them gives

$$19z = 76$$

$$z = 4$$

Now, all we have to do is to substitute the known values into the formulas and to solve for *x:*

$$4x + 5 \times 4 = 28$$

$$4x = 28 - 20 = 8$$

$$x = 2$$

and, finally, to solve for *y:*

$$x + 2y + 3z = 20$$

$$2y = 20 - 2 - 3 \times 4 = 6$$

$$y = 3$$

All three equations are satisfied when $x = 2$, $y = 3$, and $z = 4$.

A2.6 Differentiation

A2.6.1 Rule 1

$$\text{If } y = ax^n, \text{ then } \frac{dy}{dx} = nax^{n-1}$$

Differentiate:

$$y = 3x^{-3}$$

$$\frac{dy}{dx} = -3(3x^{-3-1}) = -9x^{-4}$$

$$y = \frac{x^6}{3}$$

$$\frac{dy}{dx} = \frac{6x^5}{3} = 2x^5$$

$$y = 2x^{3/4}$$

$$\frac{dy}{dx} = \frac{3}{4}(2x^{3/4-1}) = \frac{3x^{-1/4}}{2}$$

$$y = 5$$

$$y = 5x^0$$

$$\frac{dy}{dx} = 0 \times 5x^{0-1} = 0$$

Remember that anything to exponent zero is one and anything multiplied with zero is zero.

A2.6.1.1 Repeated Differentiation

When a function is differentiated, the differential coefficient is written as dy/dx or $f'(x)$. When the function is differentiated again, the second differential coefficient is d^2y/dx or $f''(x)$.

$$y = 3x^5$$

$$\frac{dy}{dx} = 15x^4$$

$$\frac{d^2y}{dx} = 60x^3$$

$$\frac{d^3y}{dx} = 180x^2$$

A2.6.2 Rule 2

A function of a function is a function in a function. If $y = (u)^v$, then $dy/dx = dy/du \times du/dx$

$$y = (2x^3 - 4x)^5$$

Let

$$(2x^3 - 4x) = u$$

Then, $y = u^5$.

$$\frac{dy}{dx} = 5(u)^4 \times (6x^2 - 4)$$

$$\frac{dy}{dx} = 5(2x^3 - 4x)^4 (6x^2 - 4)$$

A2.6.3 Rule 3

Differentiation of a product of two functions $y = uv$ is given by $dy/dx = v(du/dx) + u(dv/dx)$.

$$y = 3x^3(2x^2 - 9x + 5)^4$$

Let

$$3x^3 = u$$

and

$$(2x^2 - 9x + 5)^4 = v$$

$$\frac{dy}{dx} = [(2x^2 - 9x + 5)^4 \times 9x^2] + [3x^3(4(2x^2 - 9x + 5)^3 \times (4x - 9))]$$

$$\frac{dy}{dx} = 3x^2[((2x^2 - 9x + 5)^4 \times 3) + x((16x - 36)(2x^2 - 9x + 5)^3)]$$

$$\frac{dy}{dx} = 3x^2(2x^2 - 9x + 5)^3(3(2x^2 - 9x + 5) + (16x^2 - 36x))$$

$$\frac{dy}{dx} = 3x^2(2x^2 - 9x + 5)^3(6x^2 - 27x + 15 + 16x^2 - 36x)$$

$$\frac{dy}{dx} = 3x^2(2x^2 - 9x + 5)^3(22x^2 - 63x + 15)$$

If this looks very confusing, try another way:

$$\frac{dy}{dx} = v^4(3 \times 3x^2) + 3x^3\left(4(v)^3 \frac{dv}{dx}\right)$$

$$\frac{dy}{dx} = 3x^2(v)^3(3(v) + x(4)(4x - 9))$$

$$\frac{dy}{dx} = 3x^2(v)^3(3(2x^2 - 9x + 5) + 16x^2 - 36x)$$

$$\frac{dy}{dx} = 3x^2(2x^2 - 9x + 5)^3(6x^2 - 27x + 15 + 16x^2 - 36x)$$

This is not difficult, but the numbers can get confusing. In this problem, all three of the rules were used.

Differentiate $y = x^2(4x - 3)$.

$$\frac{dy}{dx} = 2x(4x-3)+x^2(4)$$

$$\frac{dy}{dx} = 8x^2 - 6x + 4x^2 = 12x^2 - 6x = 6x(2x-1)$$

$$y = x^2(4x-3) = 4x^3 - 3x^2$$

$$\frac{dy}{dx} = 12x^2 - 6x = 6x(2x-1)$$

This simple problem is perhaps easier to understand and verify. Now try something slightly more complicated.

A2.6.4 Rule 4

If

$$y = \frac{u}{v}$$

then

$$\frac{dy}{dx} = \frac{v\dfrac{du}{dx} - u\dfrac{dv}{dx}}{v^2}$$

Differentiate the following:

$$y = \frac{x^4}{(2x-1)^3}$$

Let

$$x^4 = u$$

and

$$(2x-1)^3 = v$$

$$\frac{du}{dx} = 4x^3 \quad \text{and} \quad \frac{dv}{dx} = 3(2x-1)^2(2) = 6(2x-1)^2$$

$$\frac{dy}{dx} = \frac{(2x-1)^3(4x^3) - x^4(6(2x-1)^2)}{((2x-1)^3)^2}$$

$$\frac{dy}{dx} = \frac{(2x-1)^2((2x-1)4x^3 - 6x^4(1))}{(2x-1)^6}$$

$$\frac{dy}{dx} = \frac{8x^4 - 4x^3 - 6x^4}{(2x-1)^4} = \frac{2x^4 - 4x^3}{(2x-1)^4} = \frac{2x^3(x-2)}{(2x-1)^4}$$

For more elaborate differential work, you should get a basic mathematics text.

A2.7 Applications of Differentiation

When it is necessary to find the turning points of a system described by a function, the first differential will give the turning points and the second differential will indicate if it is a maximum or a minimum.

The operation works as follows:

Find dy/dx and set $dy/dx = 0$ to find the turning points.

Find d^2y/dx and substitute the turning points into this second differential. If the value is positive, the point is a minimum point and, if it is negative, it is a positive point.

Find and distinguish between the maximum and minimum values of $2x^3 - 5x^2 - 4x$:

$$\frac{dy}{dx} = 6x^2 - 10x - 4 = 0$$

$$(3x+1)(2x-4) = 0$$

$$x = -\frac{1}{3}$$

$$x = 2$$

$$\frac{d^2y}{dx} = 12x - 10$$

When

$$x = -\frac{1}{3}, \quad \frac{d^2y}{dx} = 12\left(-\frac{1}{3}\right) - 10 = -14$$

this is a maximum point at

$$2x^3 - 5x^2 - 4x = 2\left(-\frac{1}{3}\right)^2 - 5\left(-\frac{1}{3}\right)^2 - 4\left(-\frac{1}{3}\right)$$

$$= -\frac{2}{27} - \frac{5}{9} + \frac{4}{3} = \frac{-2 - 15 + 36}{27} = \frac{19}{27}$$

When $x = 2$,

$$\frac{d^2y}{dx} = 12(2) - 10 = +14$$

This is a minimum point at

$$2x^3 - 5x^2 - 4x = 2(2)^3 - 5(2)^2 - 4(2) = 16 - 20 - 8 = -12$$

Appendix 3

A3.1 Metric System Steam Table

Temperature (°C)	Pressure (Absolute) (kPa)	Specific Volume v_g (m³ kg⁻¹)	Specific Enthalpy (kJ kg⁻¹)			Specific Entropy (kJkg⁻¹K⁻¹)		
			h_f	h_{fg}	h_g	s_f	s_{fg}	s_g
0	0.6108	206.3	0.0	2501.6	2501.6	0.0	9.16	9.16
2	0.6970	179.92	8.4	2496.8	2505.2	0.03	9.07	9.10
10	1.2276	106.38	42.0	2477.9	2519.9	0.15	8.75	8.90
20	2.3366	57.79	83.9	2454.3	2538.2	0.30	8.37	8.67
30	4.246	32.89	125.7	2430.7	2556.4	0.44	8.02	8.45
40	7.384	19.52	167.5	2406.9	2574.4	0.57	7.69	8.26
50	12.349	12.03	209.3	2382.9	2592.2	0.70	7.37	8.08
60	19.940	7.671	251.1	2358.6	2609.7	0.83	7.10	7.93
70	31.192	5.042	293.0	2334.0	2626.9	0.95	6.80	7.75
80	47.39	3.407	334.9	2308.8	2643.8	1.08	6.54	7.62
90	70.14	2.361	376.9	2283.2	2660.1	1.19	6.29	7.40
100	101.35	1.673	419.1	2256.9	2676.0	1.31	6.05	7.36
110	143.35	1.213	461.3	2230.0	2691.3	1.42	5.82	7.24
120	198.53	0.892	503.7	2202.2	2706.0	1.53	5.60	7.13
130	270.3	0.67	546.3	2173.6	2719.9	1.63	5.39	7.03
140	361.3	0.51	589.1	2144.0	2733.1	1.74	5.19	6.93

Temperature (°C)	Pressure (Absolute) (kPa)	Specific Volume v_g $(m^3\,kg^{-1})$	Specific Enthalpy $(kJ\,kg^{-1})$			Specific Entropy $(kJkg^{-1}K^{-1})$		
			h_f	h_{fg}	h_g	s_f	s_{fg}	s_g
150	476.0	0.39	632.1	2113.2	2745.4	1.84	4.99	6.83
175	892.4	0.22	741.1	2030.7	2771.8	2.09	4.53	6.62
200	1553.8	0.13	852.4	1938.6	2790.9	2.33	4.10	6.43
225	2550	0.08	966.9	1834.3	2801.2	2.56	3.68	6.25
250	3973	0.05	1085.8	1714.7	2800.4	2.79	3.28	6.07
300	8581	0.02	1345.1	1406.0	2751.0	3.26	2.45	5.71

Source: Adapted from Lewis, M. J. 1987. *Physical Properties of Foods and Food Processing Systems.* West Sussex: Ellis Horwood Ltd.

A3.2 Saturated Water—Temperature Table

Temp. °C	Saturated Pressure kPa	Specific Volume (m³/kg) (Saturated)		Internal Energy (kJ/kg) (Saturated)		Enthalpy (kJ/kg) (Saturated)		Entropy (kJ/kgK) (Saturated)	
		Liquid v_f	Vapor v_g	Liquid u_f	Vapor u_g	Liquid h_f	Vapor h_g	Liquid s_f	Vapor s_g
0.01	0.6113	0.001000	206.140	0.00	2375.3	0.00	2501.4	0.0000	9.1562
5	0.8721	0.001000	147.120	20.97	2382.3	20.98	2510.6	0.0761	9.0257
10	1.2276	0.001000	106.380	42.00	2389.2	42.01	2519.8	0.1510	8.9008
15	1.7051	0.001001	77.930	62.99	2396.1	62.99	2528.9	0.2245	8.7814
20	2.3390	0.001002	57.790	83.95	2402.9	83.96	2538.1	0.2966	8.6672
25	3.1690	0.001003	43.360	104.88	2409.8	104.89	2547.2	0.3674	8.5580
30	4.2460	0.001004	32.890	125.78	2416.6	125.79	2556.3	0.4369	8.4533
35	5.6280	0.001006	25.220	146.67	2423.4	146.68	2565.3	0.5053	8.3531
40	7.3840	0.001008	19.520	167.56	2430.1	167.57	2574.3	0.5725	8.2570
45	9.5930	0.001010	15.260	188.44	2436.8	188.45	2583.2	0.6387	8.1648
50	12.3490	0.001012	12.030	209.32	2443.5	209.33	2592.1	0.7038	8.0763

Temp. °C	Saturated Pressure kPa	Specific Volume (m³/kg) (Saturated)		Internal Energy (kJ/kg) (Saturated)		Enthalpy (kJ/kg) (Saturated)		Entropy (kJ/kgK) (Saturated)	
		Liquid v_f	Vapor v_g	Liquid u_f	Vapor u_g	Liquid h_f	Vapor h_g	Liquid s_f	Vapor s_g
55	15.7580	0.001015	9.568	230.21	2450.1	230.23	2600.9	0.7679	7.9913
60	19.9400	0.001017	7.671	251.11	2456.6	251.13	2609.6	0.8312	7.9096
65	25.0300	0.001020	6.197	272.02	2463.1	272.06	2618.3	0.8935	7.8310
70	31.1900	0.001023	5.042	292.95	2469.6	292.98	2626.8	0.9549	7.7553
75	38.5800	0.001026	4.131	313.90	2475.9	313.93	2643.7	1.0155	7.6824
80	47.3900	0.001029	3.407	334.86	2482.2	334.91	2635.3	1.0753	7.6122
85	57.8300	0.001033	2.828	355.84	2488.4	355.90	2651.9	1.1343	7.5445
90	70.1400	0.001036	2.361	376.85	2494.5	376.92	2660.1	1.1925	7.4791
95	84.5500	0.001040	1.982	397.88	2500.6	397.96	2668.1	1.2500	7.4159

Temp. °C	Saturated Pressure MPa	Specific Volume (m³/kg) (Saturated)		Internal Energy (kJ/kg) (Saturated)		Enthalpy (kJ/kg) (Saturated)		Entropy (kJ/kgK) (Saturated)	
		Liquid v_f	Vapor v_g	Liquid u_f	Vapor u_g	Liquid h_f	Vapor h_g	Liquid s_f	Vapor s_g
100	0.1014	0.001044	1.673	418.94	2506.5	419.04	2676.1	1.3069	7.3549
110	0.1433	0.001052	1.210	461.14	2518.1	461.30	2691.5	1.4185	7.2387
120	0.1985	0.001060	0.892	503.50	2529.3	503.71	2706.3	1.5276	7.1296
130	0.2701	0.001070	0.669	546.02	2539.9	546.31	2720.5	1.6344	7.0269
140	0.3613	0.001080	0.509	588.74	2550.0	589.13	2733.9	1.7391	6.9299
150	0.4758	0.001091	0.393	631.68	2559.5	632.20	2746.5	1.8418	6.8379
160	0.6178	0.001102	0.307	674.87	2568.4	675.55	2758.1	1.9427	6.7502
170	0.7917	0.001114	0.243	718.33	2576.5	719.21	2768.7	2.0419	6.6663
180	1.0021	0.001127	0.194	762.09	2583.7	763.22	2778.2	2.1396	6.5857
190	1.2544	0.001141	0.157	806.19	2590.0	807.62	2786.4	2.2359	6.5079
200	1.5538	0.001157	0.127	850.65	2595.3	852.45	2793.2	2.3309	6.4323
220	2.3180	0.001190	0.086	940.87	2602.4	943.62	2802.1	2.5178	6.2861
240	3.3440	0.001229	0.060	1033.21	2604.0	1037.32	2803.8	2.7015	6.1437

Temp. °C	Saturated Pressure MPa	Specific Volume (m³/kg) (Saturated)		Internal Energy (kJ/kg) (Saturated)		Enthalpy (kJ/kg) (Saturated)		Entropy (kJ/kgK) (Saturated)	
		Liquid v_f	Vapor v_g	Liquid u_f	Vapor u_g	Liquid h_f	Vapor h_g	Liquid s_f	Vapor s_g
260	4.6880	0.001276	0.042	1128.39	2599.0	1134.37	2796.9	2.8838	6.0019
280	6.4120	0.001332	0.030	1227.46	2586.1	1235.99	2779.6	3.0668	5.8571
300	8.5810	0.001404	0.022	1332.00	2563.0	1344.00	2749.0	3.2534	5.7045
320	11.2740	0.001499	0.015	1444.60	2525.5	1461.50	2700.1	3.4480	5.5362
340	14.5860	0.001638	0.011	1570.30	2464.6	1594.20	2622.0	3.6594	5.3357
360	18.6510	0.001893	0.007	1725.20	2351.5	1760.50	2481.0	3.9147	5.0526
370	21.0300	0.002213	0.005	1844.00	2228.5	1890.50	2332.1	4.1106	4.7971
374.14	22.0900	0.003155	0.003	2029.60	2029.6	2099.30	2099.30	4.4298	4.4298

A3.3 Imperial System Steam Table

Psig	Temp (°F)	Specific Volume (cu ft/lb)	Heat of the Liquid	Latent Heat (Btu/lb)	Total Heat of Steam (Btu/lb)	Psig	Temp (°F)	Specific Volume (cu ft/lb)	Heat of the Liquid	Latent Heat (Btu/lb)	Total Heat of Steam (Btu/lb)
0	212.0	26.79	180.0	970.4	1150.3	76	320.9	4.86	291.1	893.4	1184.5
1	215.3	25.23	183.4	967.2	1151.6	78	322.4	4.76	292.7	892.2	1184.9
2	218.5	23.80	186.6	966.3	1152.8	80	323.9	4.67	294.3	891.0	1185.3
3	221.5	22.53	189.6	964.3	1153.9	82	325.4	4.57	295.9	889.5	1185.7
4	224.4	21.40	192.5	962.4	1154.9	84	326.9	4.48	297.4	888.7	1186.1
5	227.2	20.38	195.3	960.4	1155.9	86	328.4	4.400	298.9	887.5	1186.4
6	229.8	19.45	198.0	958.8	1156.8	88	329.8	4.319	300.4	886.4	1186.8
7	232.4	18.61	200.6	957.2	1157.8	90	331.2	4.241	301.8	885.3	1187.1
8	234.8	17.85	203.1	955.5	1158.6	92	332.5	4.166	303.2	884.3	1187.5
9	237.1	17.14	205.4	954.0	1159.4	94	333.9	4.093	304.6	883.2	1187.8
10	239.4	16.49	207.7	952.5	1160.2	96	335.2	4.023	306.0	882.4	1188.1
11	241.6	15.89	209.9	951.1	1161.0	98	336.0	3.955	307.4	881.1	1188.5

Psig	Temp (°F)	Specific Volume (cu ft/lb)	Heat of the Liquid	Latent Heat (Btu/lb)	Total Heat of Steam (Btu/lb)	Psig	Temp (°F)	Specific Volume (cu ft/lb)	Heat of the Liquid	Latent Heat (Btu/lb)	Total Heat of Steam (Btu/lb)
12	243.7	15.34	212.1	949.6	1161.7	100	337.0	3.890	308.8	880.0	1188.8
13	245.8	14.82	214.2	948.2	1162.4	102	339.2	3.826	310.1	879.0	1189.1
14	247.8	14.33	216.2	946.8	1163.0	104	340.4	3.765	311.4	878.0	1189.4
15	249.7	13.88	218.2	945.5	1163.7	106	341.7	3.706	313.5	876.2	1189.7
16	251.6	13.45	220.1	944.2	1164.3	108	343.0	3.648	314.1	875.8	1189.9
17	253.5	13.05	222.0	942.9	1164.9	110	344.2	3.591	315.3	874.9	1190.2
18	255.3	12.68	223.9	941.6	1165.5	112	345.4	3.538	316.6	873.9	1190.5
19	257.1	12.33	225.7	940.4	1166.1	114	346.6	3.486	317.8	873.0	1190.8
20	258.8	11.99	227.4	939.3	1166.7	116	347.8	3.435	319.1	872.0	1191.1
21	260.5	11.67	229.1	938.1	1167.2	118	348.9	3.385	320.3	871.0	1191.3
22	262.1	11.38	230.8	936.9	1167.7	120	350.1	3.338	321.5	870.5	1191.6
23	263.7	11.09	232.4	935.8	1168.2	122	351.2	3.292	322.7	869.1	1191.8
24	265.3	10.82	234.0	934.8	1168.8	124	352.4	3.248	323.8	868.8	1192.1
25	266.9	10.67	235.6	933.7	1169.3	126	353.5	3.204	325.0	867.3	1192.3
26	268.3	10.32	237.2	932.5	1169.7	128	354.6	3.160	326.2	866.4	1192.6
27	269.8	10.00	238.7	931.5	1170.2	130	355.7	3.118	327.3	865.5	1192.8
28	271.3	9.86	240.1	930.5	1170.6	132	356.7	3.078	328.4	864.6	1193.0
29	272.7	9.65	241.6	929.5	1171.1	134	357.8	3.039	329.5	863.8	1193.3
30	274.1	9.45	243.0	928.5	1171.5	136	358.9	2.999	330.6	862.8	1193.5
32	276.8	9.07	245.7	926.6	1172.3	138	359.9	2.961	331.8	861.9	1193.7
34	279.4	8.72	248.4	924.7	1173.1	140	360.9	2.925	332.8	861.1	1193.9
36	281.9	8.40	251.0	922.9	1173.9	142	362.0	2.890	333.9	860.3	1194.2
38	284.3	8.10	253.5	921.1	1174.6	144	363.0	3.856	335.0	859.4	1194.4
40	286.7	7.82	255.9	919.4	1175.3	146	364.0	2.823	336.0	858.6	1194.6
42	289.0	7.56	258.3	917.6	1175.9	148	365.0	2.790	337.1	857.7	1194.8
44	291.3	7.32	260.6	916.0	1176.6	150	365.9	2.758	338.1	856.9	1195.0
46	293.5	7.09	262.9	914.3	1177.2	152	366.9	2.726	339.1	856.1	1195.2
48	295.6	6.88	265.1	912.7	1177.8	154	367.9	2.695	340.1	855.3	1195.4

Psig	Temp (°F)	Specific Volume (cu ft/lb)	Heat of the Liquid	Latent Heat (Btu/lb)	Total Heat of Steam (Btu/lb)	Psig	Temp (°F)	Specific Volume (cu ft/lb)	Heat of the Liquid	Latent Heat (Btu/lb)	Total Heat of Steam (Btu/lb)
50	297.7	6.68	267.2	911.2	1178.4	156	368.8	2.665	341.1	854.4	1195.5
52	299.7	6.50	269.3	909.6	1178.9	158	369.8	2.635	342.1	853.6	1195.7
54	301.7	6.32	271.3	908.2	1179.5	160	370.7	2.606	343.1	852.8	1195.9
56	303.6	6.14	273.3	906.7	1180.0	162	371.6	2.578	344.1	852.0	1196.1
58	305.5	5.98	275.2	905.3	1180.5	164	372.6	2.551	345.1	851.2	1196.3
60	307.3	5.83	277.1	903.9	1181.0	166	373.5	2.524	346.0	850.5	1196.5
62	309.1	5.69	279.0	902.5	1181.5	168	374.4	2.498	347.0	849.7	1196.7
64	310.9	5.56	280.8	901.2	1182.0	170	375.3	2.472	347.9	848.9	1196.8
66	312.6	5.43	282.6	999.8	1182.4	172	376.2	2.447	348.9	848.1	1197.0
68	314.4	5.30	284.4	998.5	1182.9	174	377.1	2.422	349.8	847.4	1197.2
70	316.0	5.18	286.1	897.2	1183.3	176	377.9	2.397	350.7	846.6	1197.3
72	317.7	5.07	287.8	895.9	1183.7	178	378.8	2.373	351.6	845.9	1197.5
74	319.3	4.97	289.5	894.6	1184.1						

Source: Farrall, A. W. 1979. *Food Engineering Systems,* vol. 2. Westport, CT: AVI Publishing Co.

Index

5S, 19, 20
    set in order, 19
    shine, 19
    sort, 19
    standardize, 19
    sustain, 19

A

Absolute pressure, 66, 113
Acceleration, 32
Accident prevention, 284
Activated sludge, 228
Activity efficiency, 9
Adhesion, 64
Adiabatic process, 111
Aerated lagoon, 228
Aerobacter aerogenes, 214
Air conditioning, 184
Air curtain doors, 209
Air curtains, 209
Amalgat's law, 117
Ammonia pipe, 102
Ampere, 122
Anaerobic biological, 228
Angular velocity, 88
Apparent viscosity, 61
Applications of differentiation, 319
Area, 31
Atmospheric pressure, 58, 66
Auger conveyors, 245
Automatic expansion valve, 188, 189, 192
Avogadro's number, 116

B

Back-pressure regulator, 189
Backpressure-regulating valve, 192

Base exchange resins, 221
Base units, 29
Basic causes for accidents, 282
Belt conveyors, 244
Bernoulli's equation, 77
Bilobe pump, 98
Bingham plastic fluids, 62
Bingham pseudoplastic, 63
Biological oxygen demand (BOD), 226
Bituminous coal, 124
Black body, 146
Black magnetic oxide, 224
Blasius equation, 79
Blowdown, 168
Blowdown system, 170
BOD, 226
Boiler blowdown, 168
    blowdown tank, 168
    quench tank, 168
    steam box, 168
Boiler caustic embrittlement, 169
Boiler capacity, 171
Boiler codes and standards, 170
Boiler efficiency, 166, 167
Boiler feedwater treatment, 224
Boiler fittings and accessories, 178
    fusible plug, 182
    pressure gauge, 178
    safety valve, 180
    steam pressure regulating valve, 178
    steam trap, 178
    water column tube, 179
Boiler fuel, 174
    fuel oil, 174
    gaseous fuel, 175
    solid fuel, 176
Boiler overload, 154
Boiler selection, 172

Boiler turndown, 173
Boiling, 156
Boltzmann's constant, 115
Brine systems, 202
Brundtland Commission, 275
Bucket elevators, 246
Building bylaws, 252
Building design, 251
Building fire exits, 290
Bulk hand, 247
Bulk modulus, 59

C

Cap-and-trade, 268
Capillary action, 66
Capitalism, 1
Carbon dioxide, 217
Carbon monoxide, 177
Cardinality, 25
Cation resin, 223
Causes of accidents, 282
Caustic embrittlement, 169
Cavitation, 90
Cellular layout, 14
Cellular manufacturing, 12
Centipoise, 61
Centrifugal compressors, 191
Centrifugal force, 36
Centrifugal pump
    efficiency, 89
    multistage pump, 88
    single-stage, 88
Centrifugal pumps
    operation, 89
Centrifugal pumps, 87
Chain conveyors, 245
Chlorides and sulfates, 217
Chutes, 241
Circular measure, 31
Classic suction system, 85
Cleaning-in-place (CIP), 104
Climate change, 270
Coagulants, 218
Coefficient of performance, 199
Cog V-belt, 130
Cohesion, 64
Cold lime softening, 220
Cold room, 206
Cold room doors and devices, 208
Cold room floor, 208
Cold room management, 209

Cold storage rooms, 205
Coliform bacteria, 214
Colloids, 218
Color coding, 285
Combination load, 171
Combustion efficiency, 166
Compression refrigeration sy
    compressor, 186
    condenser, 186
    evaporator, 186
    expansion valve, 186
Compression refrigeration system, 186
Compressor, 187, 189
Concave coving, 259
Concrete floor, 260
Condenser, 191
Conduction, 137
Conductors, 123
Congealing tank system, 201
Conservation of mass, 42
Controlled incineration, 231
Convection, 137
Convective and conductive heat transfer, 144
Convective heat transfer, 143
Conversion tables, 299
Conveyors, 238, 241
Corporate responsibility, 269, 277
Corporate social responsibility, 268
Corrosive gases, 224
Coulomb, 122
CPM, 2
Critical path method, 2
Culinary steam, 156, 182
Cycles, 123

D

Dalton's law, 117
Deaeration, 224
Decibel scale, 293
Decibels, 293
Defects, 10
Defrosting, 210
Density, 33, 58, 59, 61, 66, 70, 73, 74, 76, 77,
    79, 81, 82, 83, 86, 87, 106, 108, 110,
    118, 120, 143, 154, 176, 207, 218
Density, 31
Diatomic elemental gases, 116
Dichlorofluoromethane. See Freon
Differentiation, 315
Dilatant fluids, 62

Dimension, 28
   measurable extent, 28
Dimensional analysis, 32
Dimensions
   amount of a substance, 30
   electric current, 30
   length, 30
   luminous intensity, 30
   mass, 30
   temperature, 30
   time, 29
Doors, 263
Drive system management, 132
Dry steam, 150
Dust explosion, 287
Dynamic head, 91
Dynamic viscosity, 73
Dynamic viscosity, 61

E

Electrical energy, 125
Electrical motors, 125
Electrical system, 264
Electrical systems, 121
Electricity generation, 121
Electromotive force (emf), 126
Eliminate waste, 9, 13
Emergency evacuation planning, 290
Emissions control, 271
Enclosed impellers, 89
Energy, 35
Energy of a system, 109, 110
   external, 110
   internal, 110
Enthalpy, 148
Enthalpy of a system, 148
Entrained air, 92
Environmental audit
   air, 274
   recycling, 274
   solid waste, 273
   water, 274
Environmental audits, 273
Environmental compliance, 270
Environmental management system, 276
Environmental noncompliance, 270
Environmental protection agency, 252
EPA, 252
Escherichia coli, 214
Evaporating coil, 187
Expansion valve, 188

Exponents, 309
Extensive property of a system, 110
External gear pump, 97

F

Failure of pressure equipment, 286
Fanning equation, 79
Feedwater, 156
Filtration, 222
Fire fighting, 292
Fire prevention plan, 291
Fire protection, 289
Fire suppression system, 291
Fire tube boiler, 164
   efficiency, 164
   gas passes, 164
   pressure output, 164
Flammable materials, 287
Flat belt, 129
Flooded type of evaporator, 188
Floor drains, 262
Floor finishes, 259
Floor slope, 260
Floors, 259
Flow, 32
Fluid, 57
Fluidity, 60
Fluorescent lamps, 134
Foot valve, 94
Force, 32
Fork trucks, 246
Fossil fuel, 176
Foundations, 258
Fourier equation, 138, 140
Fractions, 305
Freezers, 206
Freezing, 184
Freon, 186
Frequency, 32
Frictional loss, 79
Frictional losses, 78
Fuel-to-steam efficiency, 166
Fusible plug, 182
Future climate policy, 268

G

Gas pipes, 102
Gases, 112
Gases in water, 217
Gauge pressure, 66, 114

Gear pump, 97
Gear trains, 131
Gears, 130
Gravitational constant, 32
Groundwater, 214
Gutter drains, 262

H

Hagen-Poiseuille equation, 79
Hand-washing stations, 264
Hardness in water, 215
Heat flow in a cylindrical body, 141
Heat transfer, 137
Heating load, 171
Helical gears, 131
Helical screw pump, 97
Help with calculations, 305
High-pressure boilers, 170
Horsepower, 34
Hydrogen sulfide, 218
Hydrostatic gauge pressure, 159
Hydrostatic pressure test, 159
Hydrostatics, 67
Hypoid gears, 131

I

Ideal fluid, 60
Ideal gas equation, 113
Ideal gas law, 113
Incandescent lamps, 134
Incineration, 230
Incinerator emissions, 231
Industrial water treatment, 227
Inspection of equipment, 288
Insulation, 206
Insulators, 123
Integrated system, 20, 2, 3
Intensive properties of a system, 110
Internal design pressure, 159
Internal energy, 111
Internal gear pump, 97
Inventory, 10
Inventory efficiency, 9
Ion exchange, 220
Iron in water, 216
ISO 14000, 275
ISO 14000 Series, 275
Isochoric process, 111
Isothermal process, 111

J

JIT, 19, 236
Joules, 34
Just in time, 19

K

Kanban, 4, 9, 12, 15, 236
Kilowatt, 34
Kilowatt-hour, 125
Kinds of floors, 260
Kinematic viscosity, 61
Kinetic energy, 31, 36

L

Laminar flow, 71
Landfill, 231
Landfill methods, 232
Latent heat in steam, 151
Lean 5S, 19, 20
    set in order, 19
    shine, 19
    sort, 19
    standardize, 19
    sustain, 19
Lean accounting, 8, 13, 17, 19, 21, 22
Lean Accounting, 16
Lean implementation strategy, 14
Lean manufacturing, 22, 5, 9, 13, 14, 15, 16,
    17, 18, 19, 21, 22
Lean philosophy, 8, 18
Lean production, 5, 6, 11, 12, 13, 16, 17, 18,
    21, 268
Lean thinking, 6, 7, 18, 20, 21, 22, 23
Light reflectance, 134
Lignite, 176
Load tracking, 172
Load variation, 172
Lobe pump, 97, 98
Locating the building, 253
Low-pressure boilers, 170

M

Magnetite, 225
Manganese in water, 217
Manufacturing system, 3
Mass, 30

Mass balances, 42
Mass Production, 3
Materials handling, 235
Materials handling efficiency, 236
Materials handling equipment, 241, 249
    fixed area system, 241
    fixed path conveyor, 241
    maintenance, 249
    preventative maintenance, 249
Materials handling system, 237
McDaniel Tee, 157
Mechanical refrigeration, 185
Membrane filters, 223
Metal plate floor, 260
Microbiological concerns, 214
Minimum wage, 5
Momentum, 32
Monolithic surfacing, 261
Motion, 10
Motor management, 128
Muda, 9

N

Natural refrigeration, 184
Negative index notation, 28
Net positive suction head, 91
New processing facility planning, 265
Newtonian fluids, 61
Newtons, 30, 32, 57, 59
Nitrates, 217
Nitrogen oxides, 176, 177, 274
Noise, 293
Noise control, 293
Nonrotational flow, 74
Number of boilers, 172
Number sense, 26
Numerical prefixes of SI units, 299

O

Occupational Safety and Health
    Administration, 281
Occupational, Safety and Health Act, 252
Oil in refrigerant, 188
Operating instructions, 286
Organizational metrics, 18
OSHA, 252, 281
OSHA inspection, 289
Overall heat transfer rate, 139
Overprocessing, 10
Overproduction, 10

P

Pallets, 247
Pascals, 57, 68, 87
Perfect gas law, 115
Peristaltic pump, 97, 100
PERT, 2
Phase, 123
Pipe Assembly, 103
Pipelines, 102
Piping, 263
Piping materials, 102
Piping requirements, 102
Planning the building on the site, 256
Plant safety, 287
Plastic strip curtains, 208
Pneumatic systems, 238
Pollution reduction program, 271
Poly-v-belts, 129
Portable extinguishers, 292
Portable fire extinguishers, 290
Positive displacement pump
    capacity, 96
    head, 96
    power requirement, 96
    reversing, 96
    self-priming, 96
Positive displacement pumps, 94
Potable water, 212
Potential energy, 35
Potential head, 76
Power, 34
Power transmission, 128
Prefabricated polyurethane panels, 206
Pressure, 32, 57
Pressure gauge, 178
Pressure head, 76
Pressure intensity, 67
Prevention of waste, 229
Process load, 171
Production leveling, 12
Progressive cavity pump, 101
Properties of fluid, 57
Properties of refrigerants, 192, 197
Pseudoplastic, 63
Pseudoplastic fluids, 62
Pump efficiency, 86
Pumps, 85

Q

Quadratic equations, 311

R

Radiation, 137
Radiation loss, 168
Radioactive materials, 215
Red iron oxide, 224
Refrigerant, 192
    ammonia, 193
    freon, 193
    R-125, 195
    R-134a, 194
    R-22, 196
    R-32, 196
    R-407, 195
    R-500, 195
Refrigeration and cooling, 183
Refrigeration piping, 102
Refrigeration system management, 204
Refrigerator compressors, 191
    dynamic type, 191
    positive displacement type, 191
Relative density, 59
Resistance, 122
Reynolds number, 72
Rheogram, 63
Roller conveyors, 244
Rotary pumps, 97
Rotational flow, 74

S

Safety, 283
Safety valve, 180
Salmonella, 214
Sanitation codes, 252
Saturated steam specific heat, 155
Saturated vapor, 153
Saturated vapor pressure, 152
Scalar quantities, 31
Screw conveyors, 245
Secondary refrigeration, 201
Seiketsu, 19
Seiri, 19
Seiso, 19
Seiton, 19
Semi-open impellers, 89
Seven wastes, 8, 10, 11
Shear stress, 57, 60, 61, 62, 67, 70
Shearing strain, 60, 61, 62
Shear-thickening, 63
Shear-thickening fluids, 62
Shear-thinning fluids, 62

Shear-thinning fluids, 62
Shigella, 214
Shitsuke, 19
SI system, 27, 28, 59
Simultaneous equations, 313
Single-Phase Capacitor Motor, 126
Single-phase power, 122
Single-Phase Split-Phase Motor, 125
Single-Phase Universal Motor, 125
Site selection, 253
Six Sigma, 9
Slat conveyors, 245
Softening hard water, 219
Soil-bearing capacity, 254
Solid carbon dioxide. See dry ice
Solid waste, 230
Solidus notation, 27
Solid-waste, 273
Solid-waste disposal, 229
Solution by factorization, 312
Solution using the formula, 312
Specific enthalpy, 149
Specific gravity, 59
Specific heat capacity, 150
Specific heat of ice, 184
Specific latent heat, 150
Specific latent heat of vaporization, 150
Specific mass, 59
Spur gears, 131
Squirrel-cage induction motor, 126
Stack emission, 176
    nitrogen oxides, 176
    particulate matter, 176
    sulfur oxides, 176
Stack loss, 167
Stack temperature, 167
Stainless steel pipe, 104
Standard coding scheme, 286
Standard efficiency, 9
Standard pipe, 102
Static suction head, 91
Stator winding, 127
Steady flow, 75
Steam generator, 163
Steam heating, 148
Steam piping, 158
Steam pressure regulating valve, 178
Steam purity, 182
Steam table, 154
Steam trap, 178
Stefan-Boltzmann constant, 147
Stefan's law, 146

Steradian, 31
Streptococcus pyogenes, 214
Sublimation cooling, 185
    solid carbon dioxide, 185
Sulfur oxides, 177
Sump tank, 92
Superheated steam, 151, 154
Supplementary units, 29
Supply chain, 9, 15, 273
Supply tank baffles, 92
Supporting structure, 258
Surface irrigation systems, 225
Surface tension, 64
Surface Tension, 9, 64
Suspension burner, 176
Sweet water cooling, 201

T

Taguchi, 14
Takt time, 14
Temperature of evaporation, 156
Thermal conductor, 138
Thermal efficiency, 167
Thermal insulator, 138
Thermal overload relays, 132
Thermodynamic definition of enthalpy, 148
Thermodynamics, 109
Thermodynamics of vapor compression, 196
Three-phase alternating current, 121
Thyristor, 127
Tile or paver floors, 261
Torque, 37
Toyota system, 8, 12
Transferable energy, 149
Transport, 10
Trapeze hangers, 263
Treatment of water supplies, 218
Treatment to remove turbidity, 218
Trickling filter, 228
Trilobe pump, 98
Truck service building, 254
Turbidity, 213, 218
Turbulent flow, 71

U

Uniform flow, 75
Units, 28
U-shaped cells, 12

V

Value stream, 5, 6, 8, 9, 11, 13, 19, 22
Vane impellers, 89
Vapor, 57, 112
Vapor compression cycle, 196
Vapor pressure, 66
V-belt, 129
Vector, 31
Velocity, 31
Velocity gradient, 60
Velocity head, 76, 88
Velocity of whirl, 88
Ventilation, 264
Vibratory conveyors, 245
Viscosity, 37, 38, 60, 61, 62, 63, 67, 73, 74, 78,
    82, 83, 84, 96, 98, 102, 108, 110,
    143, 174, 192
Visible light, 133
Volt, 122
Volume, 31
Volute shape, 89
Vortex, 88

W

Waiting time, 10
Walls, 259
Wastewater treatment, 225
Water column tube, 179
Water conservation, 226
Water demineralizing, 222
Water hardness, 169, 216, 220
Water impurities, 213
Water quality, 211
Water safety, 212
Water tube boiler, 165
    operational pressure, 165
    safety boiler, 165
Watt, 34, 122
Wet steam, 153
Windows, 263
Work, 34
Workplace illumination, 133
Workplace safety, 281

Printed and bound by CPI Group (UK) Ltd, Croydon, CR0 4YY

24/10/2024

01779071-0001